THE IMAGINARY MUSEUM OF MUSICAL WORKS

THE IMAGINARY MUSEUM OF MUSICAL WORKS

*An Essay in the Philosophy
of Music*

LYDIA GOEHR

CLARENDON PRESS · OXFORD
1992

Oxford University Press, Walton Street, Oxford OX2 6DP

Oxford New York Toronto
Delhi Bombay Calcutta Madras Karachi
Petaling Jaya Singapore Hong Kong Tokyo
Nairobi Dar es Salaam Cape Town
Melbourne Auckland

and associated companies in
Berlin Ibadan

Oxford is a trade mark of Oxford University Press

Published in the United States
by Oxford University Press, New York

British Library Cataloguing in Publication Data
Data available

Library of Congress Cataloging in Publication Data
Goehr, Lydia
The imaginary museum of musical works : an essay in the philosophy of music / Lydia Goehr.
p. cm.
Includes bibliographical references (p.   ) and index.
1. Music—Philosophy and aesthetics. I. Title.
ML3800.G576 1992 780'.1—dc20 91–28284
ISBN 0–19–824818–0

Typeset by Hope Services (Abingdon) Ltd.
Printed in Great Britain by
Bookcraft (Bath) Ltd, Midsomer Norton, Avon

PREFACE

What is involved in the composition, performance, and reception of classical music? What are we doing when we listen to this music seriously? Why when playing a Beethoven Sonata do performers begin with the first note indicated in the score; why don't they feel free to improvise around the Sonata's central theme? Why, finally, does it go against tradition for an audience at a concert of classical music to tap its feet? This book seeks answers to these questions as it explores the philosophical and musical ramifications of speaking about music in terms of *works*. It focuses on symphonies, concertos, and sonatas, though not on any individual works.

This book is not primarily a treatise in aesthetics, if by 'aesthetics' one means an investigation of artistic value and experience. Neither is it a history or a sociology of music, a history of the philosophy of music, nor a philosophy of the history of music, though it has much to do with all of these. It is a philosophical inquiry into how and with what effect a single concept shapes a practice. The concept is that of a musical work.

Much of the argument of the book depends upon a description of how and when this concept emerged. At the end of the eighteenth century, changes in aesthetic theory, society, and politics prompted musicians to think about music in new terms and to produce music in new ways. Musicians began to think about music as involving the creation, performance, and reception not just of music *per se*, but of works of such. The concept of a work first began to serve musical practice in its regulative capacity at this time. Musicologists and other historians of music have dated this development much earlier, usually as far back as the early sixteenth century.

If musicians before 1800 did not think about music in terms of works, how did they think about it? What conditions and thoughts characterized their musical production and how did these differ from those that characterized work-production? One may glimpse part of an answer to this question if one recognizes that musicians who produce jazz, pop songs, and many if not all types of music performed within social functions and rituals are able to manage successfully without a work-concept. Could not classical musicians have also managed thus at some stage in their history?

It is often said of philosophers of the arts that they treat all issues except those that artists and musicians really care about; that in their bid to produce theories their concerns too quickly become divorced from practical issues, especially issues having to do with the enormous and varied effect the arts have in our lives. While sympathetic to the complaint, I still think musical practice is not so ephemeral, messy, or theory-free that nothing of a theoretical nature can be said about it at all. Much of our ordinary experience of the arts presupposes an entrenched and structured, though not necessarily an explicit, understanding. How we respond to different kinds of music usually depends upon our having expectations. It is just these expectations and the understanding they entail that are here to be subjected to theory. We just have to be more careful than we apparently have been about how we theorize.

Consequently, there is in this book an overarching concern with philosophical methodology. Seeking an adequate way to theorize in philosophical terms about music, I argue for a move away from the traditional Anglo-American, analytic approach towards one rooted in history. Ultimately, the methodological argument is designed to show the ways history is indispensable to the philosophy of music and by implication to philosophy of other sorts as well.

I should like to express my profound gratitude to all those who have helped me give this book its present and final form: first and foremost, to Bernard Williams. I initially came into contact with him when he agreed to supervise my doctoral research at Cambridge University. With patience and care, he guided the direction of my thought, as well as my subsequent journey in the academic world. Sincerest thanks also go to Jerrold Levinson, who has encouraged me without limits. He gave me his own work to criticize and supported me most fair-mindedly while I did so. I was assisted at different stages by Malcolm Budd, Bernard Elevitch, Steven Gerrard, Peter Kivy, Barry Smith, Michael Tanner, and Kendall L. Walton, all of whom read earlier (sometimes very early) drafts of the manuscript or particular parts thereof; I profited greatly from their extensive criticism as well as from their friendship. I benefited also from the many conversations I had over the years with friends and colleagues from Cambridge University, the Universities of Maryland at College Park and of Nevada at Reno, and Boston and Harvard Universities. In particular, I thank Deborah Achtenberg, Amy Ayres, James Celarier, Robert

Cohen, Douglas Dempster, Mary Devereaux, Linda Gorelangton, Karey Harrison, Tim McFarland, Peter Murphy, Mark Sacks, Joseph Tolliver, Anna Wesely, and William Wilborn. This book was much improved because of their help. I am grateful to the Mellon Foundation and Harvard University for giving me the opportunity to think almost uninterruptedly for a year. Elise Springer provided me with valuable editorial assistance. Last, but not least, I am grateful to the members of the editorial staff at Oxford University Press, especially to Angela Blackburn for her patience and support, as well as for having acquired challenging comments from anonymous readers. Despite all the help I have received, however, no one except myself is responsible for any errors in the text.

I wish to give thanks of a more personal sort to my father, Alexander Goehr, my mother, Audrey Crawford, my sisters, Julia and Clare, and to the rest of my extended family. They have all encouraged me—as much intellectually and musically as emotionally. During much of the preparation of this manuscript, my husband Benjamin Kaplan assisted me in numerous ways. At the end, he helped me find the courage to put my pen down to finish the book—without doubt, the most difficult task of all. To Ben, I dedicate this book with love.

Boston, Mass. 1991

ACKNOWLEDGEMENTS

The author would like to thank the following journals for permission to use previously published material.

'Being True to the Work', *Journal of Aesthetics and Art Criticism*, 47 (1989), 55–67.

'The Power of the Podium', *Yale Review*, 79: 3 (1990), 365–81.

'Concepts, open', in H. Burkhardt and B. Smith (eds.), *Handbook of Metaphysics and Ontology* (2 vols.; Munich, 1991), i, 166–7.

The author is also grateful to Bild Archiv der Österreichischen National-bibliothek, Vienna, for permission to use on the dust-jacket a photograph of Kaspar von Zumbusch, 'Ludwig van Beethoven': bronze statuette, a minimized replica of the Viennese Beethoven-Monument; *Stagebill* magazine for permission to quote from Byron Belt's 'Concert Etiquette' printed in full in the Metropolitan Opera Program Notes.

CONTENTS

Thou Shalt Not:
Talk . . .
Hum, Sing, or Tap Fingers or Feet . . .
Rustle Thy Program . . .
Crack Thy Gum in Thy Neighbors' Ears . . .
Wear Loud-Ticking Watches or Jangle Thy Jewelry . . .
Open Cellophane-Wrapped Candies . . .
Snap Open and Close Thy Purse . . .
Sigh With Boredom . . .
Read . . .
Arrive Late or Leave Early . . .

'Concert Etiquette' by Byron Belt

Introduction

I

When E. T. A. Hoffmann had his fictional character Kapellmeister Johannes Kreisler quote an ancient law prohibiting 'noisy labourers from living next to educated gentlemen', he had a single purpose in mind. He wanted to know why 'poor oppressed composers' of his day (the early 1800s), who had to 'sell their inspirations for a price', were unable to make use of this law and 'banish themselves from the neighbourhood of windbags and bores'. Following a description of how Kreisler managed to get the better of a certain party of such windbags and bores, Hoffmann suggested that honest musicians should no longer be tortured by the extra-musical demands of social, domestic, and mundane rituals.[1]

To counter the abuse he thought music generally so subject to, and to terminate the public humiliation of musicians 'in service', Hoffmann issued an alternative prescription for musical practice. Composition, performance, reception, and evaluation should no more be guided solely by extra-musical considerations of a religious, social, or scientific sort—especially of the sort governing Kreisler's particular employment. These activities should now be guided by the musical works themselves. To legitimate this assertion Hoffmann gave currency to the notion of being true or faithful to a work (what later came to be called *Werktreue*). He gave to this notion a prominence within the language of musical thought it had never before had. 'The genuine artist,' he wrote,

lives only for the work, which he understands as the composer understood it and which he now performs. He does not make his personality count in any way. All his thoughts and actions are directed towards bringing into being all the wonderful, enchanting pictures and impressions the composer sealed in his work with magical power.[2]

[1] 'Johannes Kreislers, des Kapellmeisters, musikalische Leiden,' tr. as 'Of Kapellmeister Johannes Kreisler's Musical Sorrows' by R. Murray Schafer in *E. T. A. Hoffmann and Music* (Toronto, 1975), 121–6.
[2] 'Der echte Künstler lebt nur in dem Werke,' from Hoffmann's 'Beethovens

It is not yet necessary to describe the complex connection that exists between a musician's style of employment and the *Werktreue* ideal. For introductory purposes, the fact that Hoffmann's understanding of musical works corresponds exactly to the understanding the majority of us still have today is more pertinent. Thus most of us tend, like Hoffmann, to see works as objectified expressions of composers that prior to compositional activity did not exist. We do not treat works as objects just made or put together, like tables and chairs, but as original, unique products of a special, creative activity. We assume, further, that the tonal, rhythmic, and instrumental properties of works are constitutive of structurally integrated wholes that are symbolically represented by composers in scores. Once created, we treat works as existing after their creators have died, and whether or not they are performed or listened to at any given time. We treat them as artefacts existing in the public realm, accessible in principle to anyone who cares to listen to them. And when called, finally, to give examples of works, we usually look to the tradition of western, European, classical, 'opus' music, to works, in other words, of a 'purely instrumental' or 'absolute' sort. Who would dispute that Beethoven's symphonies, Schumann's concertos, and Schubert's sonatas are examples of musical works?

From a critical standpoint matters are rather different. Thinking about music in terms of works is not straightforward. This century has witnessed increased anxiety over the nature and implications of work-production in the field of music. This anxiety has been expressed in different ways, at least two of which have been philosophical. One of these ways begins with a 'pre-critical' description of works and ends with a description of their mode of being or ontological status given in carefully specified, metaphysical terms. But it is the intermediate reasoning that reveals the problematic nature of works, and it has usually sounded like this:

Musical works enjoy a very obscure mode of existence; they are 'ontological mutants'.[3] Works cannot, in any straightforward sense, be physical, mental, or ideal objects. They do not exist as concrete,

Instrumentalmusik', *Musikalische Novellen und Aufsätze*, i, ed. E. Istel (Regensburg, 1919), 69. For mention of the direct association between Hoffmann and *Werktreue*, see F. Blume, *Classic and Romantic Music: A Comprehensive Survey*, tr. M. D. Herter Norton (London, 1979), 112.

[3] I have borrowed this phrase from Alan Tormey. See his 'Indeterminacy and Identity in Art', *Monist*, 58 (1974), 207.

physical objects; they do not exist as private ideas existing in the mind of a composer, a performer, or a listener; neither do they exist in the eternally existing world of ideal, uncreated forms. They are not identical, furthermore, to any one of their performances. Performances take place in real time; their parts succeed one another. The temporal dimension of works is different; their parts exist simultaneously. Neither are works identical to their scores. There are properties of the former, say, expressive properties, that are not attributable to the latter. And if all copies of the score of a Beethoven Symphony are destroyed, the symphony itself does not thereby cease to exist, or so it has been argued.[4]

Even if philosophers engaged in this discussion grant that works exist apart from their performances and scores, they still maintain that both the latter stand in a peculiarly intimate relation to the work and to each other. Performances present works to us by adhering as closely as possible to the relevant scores. Our aesthetic appreciation of works is mediated through our experiences of performances and scores. One and the same work is shown to us in better and worse ways, in more or less accurate and exciting ways, through the qualitatively distinct performances we hear and the scores we read. But these assertions seem only to engender more philosophical puzzlement. What kind of existence do works enjoy, given that they are (*a*) created, (*b*) performed many times in different places, (*c*) not exhaustively captured or fixed in notational form, yet (*d*) intimately related to their performances and scores? Though not necessarily unique in their mode of existence, there is no obvious category of object within which works are comfortably placed. Just how then do they exist?

Less concerned with the ontological status of works, other theorists have given philosophical currency to the way we think about music by exploring the genesis, content, and function of musical concepts. Notable has been their concern with the many musical, cultural, and philosophical implications of speaking about music in terms of works. The German musicologists Carl Dahlhaus and Walter Wiora are among the most prominent of those who have investigated music in this way.[5]

---

[4] See R. Ingarden, *The Work of Music and the Problem of its Identity* (1928), tr. A. Czerniawski, ed. J. G. Harrell (Berkeley/Los Angeles, 1986), 1–7, for a similar introduction to the problem.

[5] See Dahlhaus, *Schoenberg and the New Music*, tr. D. Puffett and A. Clayton (Cambridge, 1987); also his *Nineteenth-Century Music*, tr. J. Bradford Robinson (Berkeley/Los Angeles, 1989); Wiora, *Das musikalische Kunstwerk* (Tutzing, 1983); and

Their investigation has been inspired by the challenge of twentieth-century, avant-garde movements to the concept of a musical work and to romantic or commodity aesthetics. Adorno has been influential here. If the concept of a work has special historical and ideological origins, this fact and those origins should be made explicit. To render the content of a concept explicit (as one does under the guise of *Begriffsgeschichte* or *Ideologiekritik*) contributes to knowledge and, when desirable, social change.

The debate over the origins of the work-concept (hereafter often referred to as such) reveals more than just an historical and political interest in music, however. It also reveals a desire to address questions about musical form and time, and about expression, contemplation, and meaning. The story of the work-concept's ascendancy has often focused on changing conceptions of form, function, and meaning.

The two philosophical approaches I have described so far make up the subject-matter of this book. The first I call *analytic*, the second, *historical*. In the first, the idea is to find the best description of the kind of *object* a work is. In the second, it is to describe the way the *concept* of a work emerged in classical music practice and how it has functioned therein. The overall argument seeks to reveal the limitations of analysis and the advantages of the historical approach.

Because I favour the historical much more than the analytic approach, my treatment of each differs. My treatment of analysis consists of a critical report of theories offered by others; that of the historical approach consists of my own version of a part of the history of music. This version rests on a controversial claim regarding the time of emergence of the work-concept, which places that time at the end of the eighteenth century. This claim quickly turns out to be the focal point for the book. It is because of this focus that the book ends up being much more than an abstract methodological inquiry into how a philosopher should best treat cultural concepts, objects, and practices.

II

The book opens by sketching a range of possible analytic positions musical works might occupy. It then looks in detail at two analytic

a more recent and very good discussion by W. Seidel in *Werk und Werkbegriff in der Musikgeschichte* (Darmstadt, 1987). For a rather different but comparable approach, see J. Attali, *Noise: The Political Economy of Music*, tr. B. Massumi (Minneapolis, 1985).

theories, by Nelson Goodman and Jerrold Levinson (Chapters 1 and 2). Details of other theories are examined *en route*. The point of looking at all these theories in detail is that a move away from analysis is justified not by prejudice, distaste, or theoretical stipulation, but because of its limitations, and the limitations of analysis are not obvious. Analysis still deserves respect, a suggestion rarely endorsed in the present climate. Thus the aim is not to dismiss analytic theories in a cursory fashion, but to dig deep in order to identify the entire range and character of questions and answers that analysts recognize when they describe the mode of being of musical works.[6]

The inquiry leads to a single, critical conclusion discussed in Chapter 3. There, it is suggested that the limitations of analysis stem from a pervasive belief (not always explicitly acknowledged) that one can arrive at an adequate philosophical understanding of what musical works are, without necessarily appealing to knowledge of how the work-concept has actually functioned in practice. This attitude of apriorism and ahistoricality manifests itself in a variety of interesting, subtle, and complex ways in the writings of analytic philosophers. My critical investigation of analysis does not presuppose, however, the correctness of any historical claim regarding the time of emergence of the work-concept. Analysis is judged as far as possible from within, by my speaking and arguing (albeit often critically) on its own terms.

That I am critical of analysis might leave readers wondering why I give so much attention to it. Why don't I just consider 'better' theories? After all, ontology can be done in ways other than analytically.[7] I have reasons.

---

[6] When I refer to analytic theories I shall be thinking especially of descriptions of musical works (sometimes part of a general description of artworks) given by N. Goodman, *Languages of Art: An Approach to a Theory of Symbols* (Oxford, 1969); P. Kivy, 'Platonism in Music: A Kind of Defense', *Grazer Philosophische Studien*, 19 (1983), 109–29; J. Levinson, 'What a Musical Work is', *Journal of Philosophy*, 77 (1980), 5–28; J. Margolis, 'Works of Art as Physically Embodied and Culturally Emergent Entities', *British Journal of Aesthetics*, 14 (1974), 187–96; R. A. Sharpe, 'Type, Token, Interpretation and Performance', *Mind*, 88 (1979), 437–40; Tormey, 'Indeterminacy and Identity in Art', 203–15; K. L. Walton, 'The Presentation and Portrayal of Sound Patterns', in J. Dancy, J. M. E. Moravscik, and C. C. W. Taylor (eds.), *Human Agency: Language, Duty, and Value. Philosophical Essays in Honour of J. O. Urmson* (Stanford, 1988), 237–57, 301–2; W. E. Webster, 'A Theory of the Compositional Work of Music', *Journal of Aesthetics and Art Criticism*, 33 (1974), 59–66; R. Wollheim, *Art and its Objects* (Harmondsworth, Middx., 1968); and N. Wolterstorff, *Works and Worlds of Art* (Oxford, 1980).

[7] See Ingarden, *The Work of Music*, J. P. Sartre, *The Psychology of Imagination*, tr. anon. (New York, 1965), 273–82; M. Heidegger, 'The Origin of the Work of Art', *Poetry, Language, Thought*, tr. A. Hofstadter (New York, 1971), 17–87; M. Dufrenne, *The Phenomenology of Aesthetic Experience*, tr. E. S. Casey *et al.* (Evanston, 1973); T. W.

First, analysis shows its strengths and weaknesses for the philosophy of music in an unadulterated way. It shows us what kind of issues we need to think about and, with its strengths and limitations, what sort of considerations we need to bring to bear on these issues. Its unadulterated character stems from its claim to be 'enlightened' and therefore uninfluenced by 'external'—sociological, political, and historical—considerations. This feigned isolation and purity ends up being a major part of the problem of analysis, but it does also render it very suitable for focused methodological assessment.

The state of analysis should not, however, be exaggerated. There is no single, definitive conception of analysis inherent in the body of aesthetical literature or in any other. The term 'analysis' functions collectively, so the most one can do is identify among its theories a criss-cross and overlap of questions and answers. Fortunately, because proponents have usually worked in response to one another, their theories constitute a body of thought in which just such an overlap exists.[8]

The most important reason I have for attending to analysis is that it has dominated the English-speaking philosophical community for several decades. In contrast to methodologies from the Continent, it has been deemed quintessentially respectable given its scientifically styled mode of argument and its high standards of clarity and internal coherence. Increasingly it has fallen into disrepute. Few theorists still seem to feel comfortable working solely within the parameters of their traditional methodology. And the number decreases as the original criticisms proffered by Continental theorists of Hegelian or pheno-menological influence are reinforced and developed by followers of Barthes, Derrida, Foucault, Gadamer, and Habermas. These attacks

---

Adorno, *Aesthetic Theory*, tr. C. Lenhardt, ed. G. Adorno and R. Teidemann (London, 1984); H. G. Gadamer, *Truth and Method* (London, 1979). Ingarden provides arguments that are remarkably similar to analytic ones. It would be useful in the following discussion, therefore, to note that much of the critical commentary applies as well to his theory as it does to analysis. I shall indicate instances where this is especially the case.

[8] Some theorists regarded here as analysts might not see themselves as such. They might deny ever having produced an analytic theory, or they might feel themselves to have already departed from their earlier analytic work. To avoid the inevitable pitfalls of using labels, I shall clarify terms like 'analysis' throughout the development of the argument. Where relevant, I shall indicate authors who, if not committing themselves to every tenet of analysis, at least endorse the claim under consideration. For recent discussion of some current attitudes towards analysis, see R. Shusterman (ed.), 'Analytic Aesthetics', *Journal of Aesthetics and Art Criticism*, 46, Special Issue (1987).

I shall presume are mostly familiar to the reader. If I do not, I shall be forced to write a general critique of analytic philosophy. Plenty of these already exist; indeed, their production has become something of a boom industry. None the less, inasmuch as this book is a methodological inquiry, it follows the prevailing spirit that seeks to comprehend analysis and then, if appropriate, to undermine its domination.

But with what precisely are we to replace analysis? However successful the critiques of analytic philosophy, the formation of an alternative has been less successful. That means I have been unable to adopt lock, stock, and barrel any single, 'new' methodology. I have had, instead, to use suggestions from many schools of thought. But I hesitate to mention them by name, lest I invite being pigeonholed into a single school by some rash reader.

Moving away from critique I begin the preferred account. This account is called historical for convenience; it might also have been called genealogy, cultural metaphysics or anthropology, or historically based ontology. The last name is most revealing because it rightly stresses that the account does not demand a complete break from ontology. To replace analysis entails looking for a new way of thinking ontologically about concepts and objects. At no point do I try to offer a complete justification for my approach (as if such a thing were possible). I never claim it is the best, or the only, alternative to analysis. To travel that route would make the book unbearably long as well as transform it into a study of pure methodology. My aim is less ambitious and more focused on musical matters. It is to show the advantages of the historical methodology by using it to treat the concept of a musical work. The proof, therefore, will be mostly in the pudding.

Chapter 4 offers the outlines of an ontology of cultural practice expressed through five claims. Articulated in musical terms, I claim that the work-concept is an *open* concept with *original* and *derivative* employment; that it is correlated to the *ideals* of musical practice; that it is a *regulative* concept; that it is *projective*; and, finally, that it is an *emergent* concept. I argue that a special connection exists between musical concepts and objects, between the way the work-concept functions and why we treat works as objects. This connection is described first as a form of projection, then historically. I describe a whole gamut of aesthetic, musical, political, and social developments first in theory (Chapters 5 and 6) and then in practice (Chapters 7 and 8) that contributed towards the founding at the end of the eighteenth century of what is well named 'the imaginary museum of musical works'—a musical institution and practice that viewed its activities and goals for the first time as conceptualized in terms of, and thus directed towards, the production and interpretation of musical works of fine art.

Central to the historical thesis is the claim that Bach did not intend to compose musical works. Only by adopting a modern perspective—a perspective foreign to Bach—would we say that he had. This implication proves to be correct as we examine with hindsight how the concepts governing musical practice before 1800 precluded the regulative function of the work-concept.

Yet nowadays we think Bach did compose works, so we have to explain that. Actually we have to explain why the work-concept has turned out to be influential in many different sorts of musical practice despite its aesthetic, musical, social, and historical particularity. Since the work-concept began to regulate practice in the romantic era, it has been employed pervasively. Philosophers and musicians alike have assumed that it can be used to speak not only about classical music but about music of almost any sort. The work-concept has found its use extended into the domains of early church music, jazz, folk, and popular music, and even into the musical domains of non-western cultures. Chapter 9 explains the extended, diversified, and sometimes hegemonic employment of the work-concept, and the tensions resulting therefrom.

In this final chapter, the model of cultural practice outlined in Chapter 4 is used to treat historical and ontological questions about conceptual use, change, and challenge. I provide reasons why some non-classical musicians have come to think of their production as work-production. I then describe this phenomenon as an example of conceptual imperialism and I connect it to an account of the diversified employment of regulative concepts. Certain aspects of avant-garde practice are then argued to constitute an attempted, but not yet altogether successful, rejection of the work-concept. I link this argument to a discussion of conceptual challenge. Finally, by pointing to the problematic status of musical professions such as conducting, and of musical movements like that of authentic performance, I point to basic problems of scope and indeterminacy in the regulative function of the work-concept.

Out of this discussion emerges the conclusion. But it is really only a question, and a broadly political one at that. It asks upon what terms, if any, we should continue to endorse the ideals associated with the work-concept, given its history and significance. 'How politically or morally bad could its history possibly have been?', readers might immediately ask themselves. 'After all, the concept only provides us with a way of thinking about music.' But this book seeks to demonstrate that thinking is never as pure or innocent as some would like it to be, and that that conclusion holds as much for our musical thought as it does for our philosophical thought.

PART I
THE ANALYTIC APPROACH

1

A Nominalist Theory of Musical Works

'The idea that music is exemplified in works . . . is far from self-evident.' This statement, made by Carl Dahlhaus, is correct.[1] Musical practice can be, but need not be, governed by the work-concept, and it is at most historically contingent whether or not it is. That a concept arises contingently out of a set of historical circumstances does not make it merely ephemeral, however. On the contrary, a concept can become so entrenched within a practice that it gradually takes on all the airs and graces of necessity. Thus it has become extraordinarily difficult for us nowadays to think about music—especially so-called classical music—in terms other than those associated with the work-concept. Yet for most of its history the tradition of 'serious' music was not thought about in these terms.

The last statement leads to the central claim of this book concerning the time of emergence of the work-concept. Its implications are immense, though that can be seen only after all the evidence has been provided. The evidence I shall provide is wide ranging yet often philosophical in nature. It ranges from an historical description of musical meaning and aesthetic theory, changes in which led to specific transformations in musical practice, to a sociological description of changing practical conditions, to an ontological description of how concepts shape practices. For strategic reasons stemming from methodological concerns, I begin with the narrowly philosophical matters, specifically the attempt by analytic philosophers to describe the ontological status of musical works.

I

What are the analytic theories of the musical work all about? Put together and scanned at a distance it is possible to identify four basic views. The next part of this chapter will briefly introduce each of these views and their primary representatives to give readers a preliminary taste of analytic theorizing.

[1] *Esthetics of Music*, tr. W. Austin (Cambridge, 1982), 10.

First there is the *Platonist* view.[2] In one of its articulations, musical works are argued, contrary to common sense, to be universals—perhaps even natural kinds—constituted by structures of sounds. They lack spatio-temporal properties and exist everlastingly. They exist long before any compositional activity has taken place and long after they perhaps have been forgotten. They exist even if no performances or score-copies are ever produced. To compose a work is less to create a kind, than it is to discover one.

This sort of view is endorsed by Nicholas Wolterstorff, though his theory is more sophisticated than the description given here. He sees works as natural kinds that, conceived as structures of sounds, are uncreated. It has always been possible that the relevant sound structures be instantiated. Composers in discovering such structures make works. They make these structures *their* works by determining for the works conditions for their correct performance. In this view, works are not just natural, uncreated kinds, but also norm kinds, because they can have properly and improperly formed examples.[3]

There is an alternative way to characterize the Platonist view. It shares with the first the idea that works exist over and above their performances and score-copies. It differs because it takes works to be quasi-Platonic entities, quasi-Platonic because they are created. Yet works retain their Platonic status because they are instantiated in performances. In this account, works are spatio-temporally bounded—dependent upon the compositional activity that brought them into existence and upon the spatio-temporal properties of particulars (performances and score-copies) that instantiate them.[4]

In Jerrold Levinson's modified Platonist view, works are structural types or kinds and their tokens are individual concrete performances. The structural types are of a specific sort; they are initiated types.

[2] Present use of Platonist (and later, Aristotelian and nominalist) terminology is standard and modern. Its use does not imply that Plato ever spoke about music in these terms. I shall also speak of score-copies rather than of the score, to remind readers that the score's ontological status is itself problematic. Is a score the sum of all score-copies, or a type comparable in ontological terms to a literary work? Does a score exist in the same way as, say, an architect's design for a building, or even a blueprint?

[3] Properties are normative within a kind if, for any norm kind, it is 'impossible that there be something which is a properly-formed example of [the kind] and lacks [the normative properties]' (*Works and Worlds of Art*, 51–7). Not all normative properties are essential, but all essential properties are normative. A good discussion of Wolterstorff's view is found in J. C. Anderson, 'Musical Kinds', *British Journal of Aesthetics*, 25 (1985), 43–9.

[4] For more on boundedness, see E. Husserl, *Experience and Judgment: Investigations in a Genealogy of Logic*, ed. L. Landgrebe (London, 1973), 264–9.

Levinson identifies a work both with a sound structure and with a performing-means structure (the instrumentation). Yet, he adds, a work, in being of a special sort, is not exhaustively identified with these two structures. It is identified, also, by its creation at a particular time, by a particular composer, in a specific historical context.[5]

One can distinguish the strong from the modified Platonist view by appealing to dependence relations. In the first view, works are independent kinds, in the second, dependent kinds. Only dependent kinds require a specific, human intentional act for them to come into existence, and performances and scores for them to stay in existence.[6] In both views, however, works are distinct entities. Distinctness is not the same as independence. To say works are distinct is to say they have existence over and above, or unexhausted by, the existence of performances and score-copies, regardless of the relation in which they stand to the latter. Distinctness characterizes the work as an abstract or concrete entity *per se*; dependence, the relation of works to their performances and score-copies.[7]

II

In an *Aristotelian* view, works are essences (typically sound structures) exhibited in performances and score-copies. As with Platonist views, works are abstract in so far as they are sound-patterns exemplified in different performances. Yet works are essential structures or patterns belonging to and inhering in other things, rather than distinct entities in their own right. Any substantiality they have is exhausted by that of their performances. This difference between Platonist and Aristotelian views is captured in Aristotle's well-known criticism of Plato's theory of forms.

---

[5] 'What a Musical Work is', 5–28.

[6] The dependence between work and performance is *virtual* in the sense that a work might not actually be being performed at a given time, though it remains a continuing practical possibility, once the kind has been created, that it could be performed. The dependence between a work and its composer is one of ontological source. Cf. Ingarden, *Time and Modes of Being*, tr. H. R. Michejda (Springfield, Ill., 1964), ch. 3 for discussion of dependence, and S. Kripke, *Naming and Necessity* (Oxford, 1980) for discussion of an essentiality of origin thesis; also C. Lord, 'A Kripkean Approach to the Identity of a Work of Art', *Journal of Aesthetics and Art Criticism*, 35 (1977), 147–55.

[7] Cf. Wolterstorff, 'Towards an Ontology of Artworks', *Noûs*, 9 (1975), 115: 'Some people will be sceptical as to whether . . . we really do have two distinct entities—a performance and that which is performed.'

[T]he most paradoxical [strange] thing of all is the statement that there are certain things besides those in the material universe, and that these are the same as sensible things except that they are eternal while the latter are perishable. For they say there is a man-himself and a horse-itself and health-itself, with no further qualification . . . a procedure like that of the people who said there are gods, but in human form. For they were positing nothing but eternal men, nor are the Platonists making the forms anything other than eternal sensible things.[8]

To obfuscate matters, Aristotle spoke of essences as having secondary existence derivative of the primary existence of concreta embodying them. Perhaps, then, one should speak of works as being distinct in a secondary sense, just because they are the essences multiply embodied in performances and score-copies. To speak in this way would allow for the view that essences might exist independently in a secondary sense as well, perhaps as unactualized or unsubstantiated potential.

Kendall Walton's account is Aristotelian in so far as he identifies each work with the sound pattern presented in each and every one of its performances. In his view, a work is something that is not itself a 'particular object or event' (a substantial thing), yet is such that performances may 'fit' or fail to fit it. A work, he argues, is a hierarchically structured pattern of sounds and instrumental specifications recognized and specified by the score, '*minus* whatever advice for good performance it contains'. It is the pattern of sound properties a performance must exhibit if the performance is to be a correct or flawless performance of the work. Once gain, works come close to being regarded normatively or prescriptively—as recipes, algorithms, or designs.[9]

III

The third way to conceive of works is to attribute to them no form of abstract existence. To talk of works is to talk only *as if* there were works; only concrete performances and score-copies exist. Works are no more than extensionally defined classes of performances-of-a-work, where '-of-a-work' is treated as a syncategorematic predicate.

---

[8] *Metaphysics*, III. 2: 997b5–10 (*The Basic Works of Aristotle*, ed. R. McKeon (New York, 1941), 720).

[9] 'The Presentation and Portrayal of Sound Patterns,' 237–57. Cf. W. Charlton's view of works as invented possibilities fulfilled in performances, in *Aesthetics* (London, 1970), 27–33, and K. Price's view of works as bundles of essences, in 'What is a Piece of Music?', *British Journal of Aesthetics*, 22 (1982), 322–36.

Performances are classified, not because they instantiate or embody abstract patterns, but because they stand in appropriate relations to one another and to score-copies. Hence, works are no more than linguistic items—general names or descriptions—serving as convenient ways to refer to certain classes of particulars, just as a surname conveniently picks out the members of a family-class who are biologically or legally related. This characterization falls under the *nominalist* view. Here, one moves away from considering the *vertical* relation between a work and its performances, a relation obtaining between an abstractum and its concreta. One considers, instead, the *horizontal* relations obtaining between performances and score-copies.

Something like this view is advocated by theorists who, following in the Peircean tradition, choose to call works *types*, or, more carefully, choose to call work-names types. A musical work-name stands to its performances as a type stands to its tokens. The type does not exist other than linguistically, even though the corresponding tokens are identified in relation to the type. In creating a type a composer creates no more than a token-of-a-type or the means for producing a token-of-a-type, i.e. a score-copy.

This is almost the position held by Joseph Margolis. He moves away from strict nominalism, however, because he sees types as 'abstract particulars'. His view comes close to being a modified Platonist view, despite his remarking upon 'the extreme implausibility of platonizing with respect to art'. An abstract particular is abstract because it can be instantiated, and it is a particular because, in being created, it is not a universal. (Hence, he is not a strong Platonist.) A work is embodied in physical objects to which it is not identical. The 'is' of embodiment, he argues, is not the same as the 'is' of identity. The former is introduced in cultural contexts to 'facilitate our discourse about cultural particulars'. The work-type, he finally proposes, is a culturally emergent entity, emergent in the sense of its being embodied in particular objects within a defined cultural space.[10]

---

[10] 'Works of Art as Physically Embodied and Culturally Emergent Entities'; Cf. Webster's view of works as abstract particulars in his 'A theory of the Compositional Work of Music', 60–1; also P. F. Strawson's description of a type as 'something which, while not itself a particular physical phenomenon, can be embodied on different occasions in different physical phenomena' ('Aesthetic Appraisal and Works of Art', *Freedom and Resentment and Other Essays* (London, 1974), 183). For a commentary on type–token views see N. Harrison, 'Types, Tokens and the Identity of the Musical Work', *British Journal of Aesthetics*, 15 (1975), 336–46. A comparable but little-known discussion of 'types' is to be found in Husserl's piece on 'bound idealities' (*Experience and Judgment*, 264–9).

Nelson Goodman is more committed to nominalism. He proposes that works are classes of performances that are perfectly compliant with scores. His theory implies no extra commitment either to a distinct or an abstract entity called a work. A commitment to classes means a thoroughly nominalist account is not provided; classes are abstract entities. Notwithstanding, Goodman claims his use of the terminology of classes is a convenience and that the substance of his theory is truly nominalistic.[11]

IV

Another way to think about works originates in the writings of Benedetto Croce and R. G. Collingwood.[12] Neither theorist produced an analytic theory of works; on the contrary. Nevertheless, one can extrapolate from their writings what has come to be known as an *idealist* view. Works are now identified with ideas formed in the mind of composers. These ideas, once formed, find objectified expression through score-copies or performances and are, thereby, made publicly accessible. Works are not identified with the objectified expressions, as one might expect them to be, but with the ideas themselves.

Collingwood's view was more complicated, as of course was Croce's. Collingwood explains:

When a man makes up a tune, he may . . . hum it or sing it or play it on an instrument. He may do none of these things, but write it on paper . . . [H]e may do these things in public, so that this tune . . . becomes public property . . . But all these [activities] are accessories of the real work . . . The actual making of the tune is something that goes on in his head, and nowhere else.[13]

[11] *Languages of Art*, p. xiii. On p. 131, Goodman also writes: 'The type is the universal or class of which marks are instances or members. Although I speak . . . of a character as a class of marks, this is for me informal parlance admissible only because it can readily be translated into more acceptable language. I prefer . . . to dismiss the type altogether and treat the so-called tokens of a type as *replicas* of one another.' Note that, in this whole body of literature, type–token terminology is inconsistently employed and is done so by nominalists and Platonists regardless of differences in ontological commitment. Cf. R. R. Dipert's insightful comment that 'there is hardly any connection at all, beyond the use of words, between the recent uses of the notions of token and type in music and the only adequate theory of what these terms mean, and the only account most authors cite, namely, C. S. Peirce's' ('Types and Tokens: A Reply to Sharpe', *Mind*, 89 (1980), 588).

[12] Croce, *Aesthetic as Science of Expression and General Linguistic*, tr. D. Ainslie (New York, 1922), and Collingwood, *The Principles of Art* (Oxford, 1938).

[13] *The Principles of Art*, 134. For discussion of idealist views, see R. Hoffman, 'Conjectures and Refutations on the Ontological Status of the Work of Art', *Mind*, 71 (1962), 512–20.

That thing existing in a person's head 'and nowhere else' is 'an imaginary thing', Collingwood continues. The making of the tune is the making of an imaginary tune. This is a case of creation. But when someone writes down their tune, to serve the purpose of informing others or perhaps of reminding themselves of the imaginary thing, this is not, Collingwood says, 'creating at all'. It is fabrication. Tunes exists only within the context of imaginative experiences. In each such experience a tune is imagined but not, literally speaking, heard. But it would be wrong to say that tunes are just these imagined sounds. They are rather the whole experience of the total activity of imagining, apprehending, and being made conscious of these sounds.[14]

Idealist theories have not been well received in the analytic tradition. The idea of identifying a work within the context of an imaginative or aesthetic experience has been judged to conform least well to the kind of ontological theory sought after. Equating the total imaginative experience with a mental entity—an idea—has not helped either. Identifying works with anything existing in the mind has been regarded a most unsatisfactory manoeuvre.[15]

V

Even from a cursory glance at these four basic positions on the status of works, one can extrapolate quite a lot about the nature both of analytic theorizing and musical works themselves. Analysis, first, shows a basic concern to describe the mode of existence of works in terms of universals, types, or kinds. As one would expect this concern is no simple matter. It carries with it a mountain of metaphysical baggage. Related to this concern is another and it, too, is complicated. It is to determine the identity and individuation conditions for works. What conditions have to be met for something to be a musical work? How is its identity over time explained? What are the individuation conditions which differentiate one work from another? Influenced by Occamist and reductionist tendencies, analysts have found a major point of contention to be the extent of their ontological commitment. In speaking of musical works is a commitment to abstracta necessary?

[14] Collingwood, *The Principles of Art*, 133–51. Collingwood speaks of tunes, but his points clearly apply to musical works as well.
[15] See Levinson, 'What a Musical Work is', 5; and his comment (loc. cit.) that the Crocean view 'puts the objectivity of musical and literary works in dire peril'. Cf. also Wollheim, *Art and its Objects*, 21.

From a musical perspective, analytic theories have commented on the role creativity and compositional activity play in determining the nature of musical works, on the role played by a composer's instrumental specifications (which notes are played on which instruments), and on the relations performances and scores have to the work and to each other. A central feature of the work-concept is immediately exposed. A work is not generally thought to be just any group of sounds, but a complex structure of sounds related in some important way to a composer, a score, and a given class of performances. To understand the idea of a musical work is to understand all these elements in their interrelations.

Taking a step back, it is not unreasonable to think that a successful theory is one that reconciles philosophical with musical interests, one that balances a philosophical understanding, say, of creation, with an understanding of what it means to compose a work. It is one sensitive to the roles the work-concept has (and does not have) in practice. It is one that explains what sort of thing a work is in terms compatible with the description of how the concept of a work functions in practice. It would demonstrate that thinking in particular theoretical terms enables one to illuminate the phenomena one wants to account for, in the way one wants to account for them. From this vantage-point one can now ask whether a successful analytic theory has been produced.

To answer this question I shall look in the rest of this chapter and in the next at two accounts—Nelson Goodman's and Jerrold Levinson's— that differ from one another significantly. Both are idiosyncratic, though still representative of analysis. I begin with Goodman's nominalist account because it makes fewer ontological commitments than Levinson's. (Levinson also produced his theory partially in reaction to Goodman's.) Given various criticisms of Goodman's theory, the motivation for making more commitments is rendered transparent.

Goodman offers an argument for the central role played by the score and performances in determining the status and identity of works; Levinson places emphasis, also, on the role played by composers in creating their works. Goodman's interest is in philosophical theory, such that features of musical works are appealed to to meet specific theoretical demands. His account has struck most theorists as counter-intuitive. Levinson tries to accommodate more pre-critical intuitions; he tries to balance his interest in philosophical theory with our more commonplace understanding of musical phenomena. To my mind neither account is successful. It is this conclusion that prompts one to ask whether equilibrium between philosophical theory and musical practice is possible at all while working within the parameters of analytic philosophy. I shall argue that it is not.

VI

In his *Languages of Art* Goodman aims to provide a general, *extensionalist* theory of symbols, where a symbol is conceived in a 'very general and colourless' way to comprehend letters, words, texts, pictures, and models. On the way he proposes a theory of the identity of musical works. He wants to show that works can be conceived as symbol systems, and that this conception explains the primary function of a score and how performances are classified as being performances of a given work. Since the particulars of his theory of works have been laid out and discussed extensively in the philosophical literature, I shall focus only on those aspects of his theory relevant to the argument of this book.

The distinguishing feature of a work lies not in there being some *abstract* entity existing apart from its performances and score-copies, but in there being for each work a special kind of score. Goodman calls it a uniquely specified character in a notational system. The special role played by notation is characteristic of all those arts that are, in Goodman's terms, *allographic* as opposed to *autographic*. An allographic art—typically an art instantiated in performances or readings—is, unlike an autographic art, one for which the distinction between original and forged instances of the art work is insignificant because 'no historical information concerning the production of the performance can affect the result' (118).[16] Any instance of an allographic art work, however produced, counts as a genuine instance, if and only if it is perfectly compliant with a carefully specified notational character—a score or a text. 'Where the works are transitory,' Goodman writes, 'as in singing or reciting, or require many persons for their production . . . a notation may be devised in order to transcend the limitations of time and the individual. This involves establishing a distinction between the constitutive and contingent properties of a work' (121).

Notation is central to those transitory works presented on a number of different occasions, to works that are ephemeral. It is central to

[16] For the rest of this chapter, unless otherwise indicated, all parenthetical page references will be to Goodman's *Languages of Art*.

works that cannot be produced in their entirety by a single person. Musical works seems to fulfil these conditions, but does it then follow that their *identity* is to be explained in terms of this notation? Goodman thinks so. A work's identity is retained in its performances, or performances are classified as being *of* a given work, just because the sounds indicated in the score are complied with in the performances. Any other, non-notational feature is irrelevant in determining the individuation of the performance and, thereby, the identity of the work itself (128–9).

What is a musical work conceived as a class of performances compliant with a score? A score is a character written in a notational system. A notational system like any language consists of characters correlated with a field of reference. Each character is a class of inscriptions (marks or utterances). A notational system, specifically a musical system or language, consists of scores (perhaps only one) correlated with classes of performances. For each work there is a single score correlated with a single class of performances. The relation specified by the correlation between score and performance is that of *compliance*. It is a one-directional semantic relation binding compliants to character-inscriptions, binding performances to scores.[17] A musical work can be represented accordingly:

score-copy (of **W**)　　　　　score-copy (of **W**)　　　　　. . .

performance (of **W**)　　　　performance (of **W**)

['(of **W**)/ means 'of a work'; the clause is bracketed to indicate syncategorematic status.]

A notational language consists in *atomic* characters which in their modes of combination form *compound* characters of greater and lesser complexity. An atomic character is formed by exactly one single character; any other is compound. Pitch characters are atomic characters in musical scores. The constituents of compound characters stand to one another in modes of combination—in relations 'prescribed by the governing rules of combination', rules governing, say, harmonic,

---

[17] 'Compliance' is interchangeable with 'detonation' or 'extension'. Performances make up the extension class of the score (144). Goodman also distinguishes a notational system from a notational scheme (130–1). In the latter, characters are correlated with a field of reference, i.e. semantically determined; in the former, they are understood syntactically. The distinction is not of particular concern here.

rhythmic, chordal, and intervallic sequences (142). The compounded-ness of characters has no upper limit, which means that a score itself can be conceived as a compound character.

By recording the constitutive features of a work in all their compoundedness, the score preserves the work's identity. It is able to serve this latter purpose, however, only because of the peculiar nature of the notational language in which these features are specified, or so Goodman argues. For a language to be 'notational' in Goodman's sense, it has to meet five syntactic and semantic requirements, requirements which effectively forbid ambiguity, overlap, and indeterminacy in and among characters and compliants (186–7). In brief, these requirements are:

(i) *Syntactic disjointness.* Characters must be disjoint so that no inscription belongs to more than one character. All inscriptions must be syntactically equivalent, intersubstitutable without syntactic effect. This is guaranteed among a character's inscriptions if each inscription is a 'replica' of all the others. These inscriptions must, therefore, also be 'character indifferent'. Character indifference is a reflexive, symmetric, and transitive relation which, by obtaining, produces a class of character-indifferent inscriptions under the partition generated by this relation (131).

(ii) *Syntactic differentiation.* Characters must be finitely differentiated: 'For every two characters K and K′ and every mark [inscription] M that does not actually belong to both, determination either that M does not belong to K or that M does not belong to K′ is theoretically possible' (135–6).

(iii) *Unique determination.* Each character must uniquely determine an extension, the membership of which is invariant over context. Thus ambiguity of inscriptions is forbidden (148).

(iv) *Semantic disjointness.* Compliance classes must be disjoint; no intersection of compliance classes is admissible (150–1); finally,

(v) *Semantic differentiation.* Given a compliant, it must be sufficiently differentiated from any other so that determination that it complies with the character in question is possible (15).

According to Goodman, the primary theoretical function of the score is served when these requirements are met. If the score is notational in his sense, one can identify the same work in each and every score-copy and performance indefinitely often. In specifying these requirements we are, in other words, providing a decisive test for determining whether a work's identity is preserved in a set of score-copies and performances. I shall refer to this as *the retrievability test*. Given a score-copy, it is possible to identify the constitutive properties of the relevant work and thereby of its performances. It is also possible to retrieve the score on hearing a performance (128). The identification procedure functions in both directions.

I shall consider Goodman's views on compliant performances in a while. Initially, I shall focus on aspects of his theory having to do with the score. The purpose of the following examination is to reveal Goodman's mode of argument, by recording what he actually says, by noting critical objections that have been and could be put to him, and by noting what his responses have been and would be likely to be. All this information, in addition to that provided in the subsequent chapter, will then be used as the foundation for a more general investigation of analytic theorizing.

VII

How well does the language of musical scores meet the notational requirements? According to Goodman, it 'is clear' that our musical scores satisfy at least his syntactic requirements. A note-mark, he writes, 'may, indeed, be so placed that we are in doubt about whether it belongs to one note-character or another, but in no case does it belong to both'. Still, he concludes, 'most characters of a musical score . . . are syntactically disjoint and differentiated. The symbol scheme is thus substantially notational' (181).

What about tempi, dynamic, and timbre markings? Do these meet any or all of Goodman's conditions? Tempi specifications are notoriously vague. Goodman is able to protect himself, and indeed does so, against all such examples. He argues that since tempi markings and the like are not notational (they do not meet the requirements), they do not count as constitutive properties of a work. Performances 'may differ appreciably in such musical features as tempo, timbre, phrasing and expressiveness', he reminds us, but they 'cannot be accounted integral parts of the defining score'. Rather, they are auxiliary directions 'whose observance or nonobservance affects the quality of the performance but not the identity of the work' (117 and 185). If certain specifications affect only the *quality* of a work, they cannot be used to challenge a description of the *identity* of a work, since issues having to do with quality are distinct from those having to do with identity, or so Goodman believes.[18]

What about cases involving syntactic difference but semantic identity? The augmented second and the minor third have identical extension despite their syntactic difference. What about scores that refer to transposing instruments? Here the mark on the page does not uniquely determine a compliance class. William Webster argues that Goodman's semantic conditions are 'violated every time a transposing instrument and a non-transposing instrument correctly read the same notational scheme in unison'.[19] Goodman concedes there are ambiguous and problematic characters in scores (181). None the less, he says, the score

comes as near to meeting the theoretical requirements for notationality as might reasonably be expected of any traditional system in constant actual use, and that the excisions and revisions needed to correct any infractions are rather plain and local. After all, one hardly expects chemical purity outside the laboratory. (186)

Goodman's critics find these 'excisions and revisions' to be neither plain nor local, and subsequently judge his theory to be unsatisfactory. Goodman draws a different conclusion, for his aim, he says, is only to specify the conditions a score must meet *in theory*. He speaks of what is theoretically possible or decisive, and what must theoretically be the case.[20] Under what conditions and circumstances scores function outside the theoretical laboratory is not his concern. That their functioning in practice is far more untidy than his theoretical account would have us believe is not his problem. His concern is to produce a coherent theoretical account, its agreement with practice being a different and difficult issue. His opponents, by contrast, judge his theory less on purely theoretical grounds than on the basis of its

[18] '[E]ven where the constitutive properties of a work are clearly distinguished by means of a notation, they cannot be identified with the aesthetic properties' (120). In much of the current discussion, the term 'aesthetic' is used to denote aspects of works having to do with quality and value, and, as such, its use is contrasted with the classificatory use of the term 'identity'. As we shall soon see, this contrast is problematic.

[19] 'Music is not a Notational System', *Journal of Aesthetics and Art Criticism*, 29 (1971), 495.

[20] '[T]o take notation as . . . nothing but a practical aid to production is to miss its fundamental theoretical role' (127–8). A few lines earlier Goodman speaks of the essential role of the score, and in this context, 'essential' is interchangeable with 'theoretical'. On p. 128 he speaks of theoretic properties. On p. 213, when considering the possibility of a notational system for dance, he writes: 'Practical feasibility is another matter, not directly in question here.'

agreement with practice. Could Goodman ever satisfy his critics? Here we see the beginnings of a critical disagreement.

VIII

In the debate over the identity of musical works a danger has usually arisen for any theorist who specifies, in a general and fixed way, certain features as constitutive of a work. This danger arises less when one tries to specify for a given work which properties are constitutive and which contingent, than when one attempts to make a general specification once and for all, for any and every musical work. To produce this sort of description is to assume there is only one kind of musical language in which works can be written. It is highly likely, though, that for each general description someone will mention examples we *think* are musical works which fail to fit it. This happened. Paul Ziff used Tartini's 'Devil's Trill' Sonata as a counter-example to Goodman's theory. Ziff argued that given Goodman's notational requirements, the essential trills in the sonata have to be regarded merely as contingent. This is a problem for Goodman if Ziff's argument is valid, but is it?[21]

Unlike Goodman, Ziff presupposes a connection to exist between issues of identity and quality. If a property is essential to the work's identity, it must be exemplified in every performance, and this affects the work's quality and character. Ziff might well want the relation to hold also in the opposite direction. If, for the sake of quality, one wants to hear the trills played in every performance of Tartini's Sonata, the trills should be scored so that they count as constitutive of the sonata's identity. This move is not as simple as it seems but, for the moment, assume that it is plausible to think that the trills specified in the sonata are essential to its identity as well as its quality. Now Ziff's challenge against Goodman still holds, or does it?

Upon what grounds can the force of Ziff's challenge now be judged? Might Goodman say Tartini's Sonata is one of those difficult cases that does not fit his theory, but a theory is still acceptable if it fits most cases or at least the standard ones? There are always difficult examples because there is no chemical purity outside the laboratory.

---

[21] 'Goodman's *Languages of Art*', *Philosophical Review*, 80 (1971), 514. The title otherwise given to Tartini's Sonata is the Sonata in G Minor for violin, 'The Devil's Sonata'. For the story behind the name see E. Heron-Allen, 'Tartini', in G. Grove, *A Dictionary of Music and Musicians*, 5th edn., ed. E. Blom, viii (New York, 1954), 313.

There is an inevitable mismatch between formal theory and informal (real) examples. Responses like these are compatible with Goodman's claim that musical scores are 'substantially' notational. Were Goodman to respond in this way, he would still have to explain, however, why the case of Tartini's Sonata is a difficult one, or allay Ziff's worries in some other way. He might suggest that for certain works, trills are counted as constitutive properties, and Tartini's Sonata is one of them. But upon what would this judgement be based?

Character signs for trills are, in the musical language used by Tartini, shorthand signs for complicated patterns of notes; they are complex characters conforming to Goodman's notational specifications. Tartini did not have to write his trills out in full (though he might still have done), because he used a language typical of the eighteenth century that incorporated over a hundred kinds of ornamentation and embellishment, each of which was assigned a notational symbol. Often these trills were differentiated and specified. Sometimes, where particular trills were neither specified fully nor indicated by general markings, embellishments were performed according to general Baroque principles of 'necessity, economy, uniformity, variety, and suitability', and, if not according to these principles, then according to conventions adaptable to the performer's taste. Rarely if ever did musicians consider the use of trills an optional matter. On the contrary, Bénigne de Bacilly wrote in 1668, the trill is 'one of the most important ornaments, without which the melody is very imperfect . . . [A]lthough the composer has not marked on paper the joins and the trills . . . it is a general rule to assume them, and never to suppress them'.[22] And regarding all smaller ornaments, Friedrich Wilhelm Marpurg wrote nearly a century later, 'they are so essential at most places that without their strict observance no composition can please the more refined ears.' But where, he continues, 'does one learn what notes are given ornaments or at which point of the melody this or that ornament ought to be introduced?' His answer was that 'one should hear persons who are reputed to play elegantly, and one should hear them in pieces one already knows. This way one may form one's taste, and do likewise.

---

[22] R. Donnington, *Baroque Music: Style and Performance: A Handbook* (New York, 1982), 126. On the 18th-cent. equation of taste and embellishment, see C. Burney, *An Eighteenth-Century Musical Tour in France and Italy* (1773), ed. P. A. Scholes (Westport, Conn., 1979), p. xxxiv: ' "Taste"—the adding, diminishing, or changing a melody, or passage, with judgment and propriety, and in such a manner as to *improve* it; if this were rendered an invariant rule in what is commonly called *gracing*, the passages, in compositions of the first class, would seldom be changed.'

For it is impossible', he remind us, 'to devise rules to meet all possible cases.'[23]

Ziff uses Tartini's Sonata (as well as other examples of early music[24]) to challenge Goodman's theory. Is his use of these examples legitimate? Is there something to be learned from the fact that all the examples Ziff uses were composed in a period when musicians did not use a modern, fully specifying notation? Has Ziff overlooked an important development in the eighteenth century in the idea of notational precision?

IX

In order to avoid the consequences of what we nowadays speak of as a (more) minimal notation, eighteenth-century composers increasingly specified in their score what they wanted to hear in performances. Many characters (and not only ornamentations) once notationally unspecified were changed into characters to be specified. 'The impromptu figuration of today becomes the written figuration of tomorrow.'[25] In 1722, Couperin illustrated the beginnings of this move from a more *skeletal* to a *through-composed* music when he expressed his surprise, 'especially after the pains I gave myself to mark the ornaments which suit my Pieces . . . to hear of people who treat them without respecting them. This is an unpardonable neglect, in view of the fact that it is not at all an optional matter to take such ornaments as one wishes.'[26] Some musicians resented the change. Commenting negatively upon J. S. Bach's use of ornament and detailed scoring, Johann Adolf Scheibe wrote: 'All the graces, all the embellishments, everything that is ordinarily taken for granted in the method of performance, he writes out in exact notes, which not only deprives his pieces of the beauty of harmony, but makes the melody totally indistinct.'[27]

[23] Donnington, *Baroque Music*, 109.

[24] For the most part I shall refer to music composed before the late 18th cent. as early music. It must not be assumed, however, that all such music is homogeneous. The name 'early music' merely serves to indicate a contrast that becomes increasingly crucial in this book, between early and 'modern', post-1800, opus music.

[25] From Donnington, 'Ornamentation' in Grove, *Dictionary*, 5th edn., vi. 374.

[26] Donnington, *Baroque Music*, 108.

[27] From Scheibe's 'Letter from an Able Musikant Abroad', 14 May 1737, in Donnington, 'Ornamentation', 375. The letter formed part of a larger correspondence recording a debate between Scheibe and J. A. Birnbaum over Bach's musical talents. Their disagreement began over an ambiguous play on the then common word '*Musikant*' which originally meant a musician, but which gradually took on the derogatory connotation of a 'fiddler'. See H. T. David and A. Mendel (eds.), *The Bach Reader: A Life of Johann Sebastian Bach in Letters and Documents* (New York, 1966), 237–8.

The change looked inevitable, however. The need for a new kind of notational precision largely resulted from a novel desire to preserve the identity of music exactly as determined by the composer. With ever more precise notation it was possible to produce numerous performances of the same sound and instrumental specifications; it became increasingly easy to hear and recognize exactly the same piece of music in different performances.

As composers came to have more control over the creation and production of their music, as they came to conceive of their music as being preservable in fixed and lasting works, they gradually took control of their performances. The nineteenth-century critic for the *Daily News*, George Hogarth, reported that in 1814, after failing to recognize an aria from his own opera, Rossini began to write out his embellishments in full to prevent them, as Hogarth put it, 'from being *disfigured* by the presumption and bad taste of the singers'.[28] Disfigurement of a composer's music was to be expected if performers had only minimal specifications to follow.

Composers' attitudes were also gradually changing with regard to instrumental specifications. These specifications, indicating which note was played by which instrument, were neither always made explicit in the score nor always considered essential to the music's identity. Much music was composed to suit, as Couperin tells us, 'not only the harpsichord, but also the violin, the flute, the oboe, the viol and the bassoon'.[29] Gradually, however, increasingly strict specifications were given for the kind and the number of instruments to be played in the performance of each and every musical work.

Generally, there was a simple remedy for dissatisfied composers. If composers wanted performers to regard certain aspects as indispensable to the performance of their compositions, they should specify these aspects more precisely and performers should learn how to follow the specifications.

X

It is not yet necessary to make the significance of the historical data fully explicit, because the relevant point is already clear. The data

---

[28] *Music History, Biography and Criticism* (New York, 1845), 153. The point is also discussed by Stendhal, *The Life of Rossini*, tr. R. N. Coe (London/New York, 1985), 353–5.

[29] Donnington, *Baroque Music*, 167.

reveals a change in the language of musical notation and this has bearing on the quarrel between Goodman and Ziff. Ziff wants to claim that the trills of Tartini's Sonata are essential. He has history on his side. The historical remarks show that not everything essential to a musical performance was always written out in full. At a given time there were unwritten rules concerning the way in which minimally notated scores determined their compliance classes. Herein lies the crux of the problem.

Goodman is likely to maintain that despite the history of scoring practices, many early music scores do not meet his notational requirements. The basic problem with minimally notated scores is that there are overlaps in their compliance classes and this is a serious failing. Just as persons cannot share bodily parts without that threatening their individual identities, so musical scores must be uniquely specifying if each work's identity is to be retained (183).

Suppose to overcome the problem we translated (through standard modernizing or editing techniques) the original score of Tartini's Sonata into a modern, 'notational' score. Would the problem now be solved? Were we to write Tartini's trills out in full no one would regard them as contingent properties and no one would choose this sonata as a counter-example. This solution can be generalized to cover any and all vague musical elements. All we have to do is translate our 'inadequate' scores into notationally 'adequate' ones.

This solution accepted, Goodman's theory would now be threatened only if it depended upon a fixed and general specification as to what counted as the notational characters in a musical language and if this specification excluded all newly translated characters. It does not though; his theory is unaffected by admitting the existence of more than one musical language.[30] That there might be and are numerous musical languages carries no theoretical weight. When composers specify constitutive properties for their works, they do so by choosing certain musical elements as constitutive. Typically they indicate these in scores. So long as scores are notational in Goodman's strict sense, they can serve the function he conceives for them. Technically this solution works even if musically it offends those concerned, say, to preserve the original character of early music scores. The question now is whether it is the best theoretical solution.

---

[30] Goodman makes numerous references to different musical languages (179 and 214 ff.).

XI

Goodman can avoid answering this last question because he does not opt for a solution involving translating 'inadequate' into 'adequate' scores. He opts for an alternative solution. Recall the historical data which showed that not all performances have always come about by virtue of a performer's following a fully specifying score. Many performances of music involve improvisation; many presuppose that performers will embellish and follow general principles associated with genre and occasion—in fact most musical performances do so, especially of a non-classical kind. The question now arises how closely the concept of a work of music, in contrast to that of music more broadly conceived, is related to the idea that music be fully specified and multiply exemplified in performances. If we are dealing with a single musical performance, say, when musicians freely improvise, how we preserve the identity of a fully composed work in a succession of performances is not an issue. Perhaps not all music is to be thought about in terms of works.

Goodman's concern is not with the identity of music broadly conceived. It is only with how the identity of works, whatever they are, can be preserved in a string of performances. His theory relies on the idea that performances of works (and not any other kind of musical performance) acquire their identity in virtue of the relation in which they stand to their scores. And these scores serve as the authoritative principle of identification for works and their performances if and only if they are adequately notated.

Perhaps when dealing with music for which, and for whatever reason, there is no adequate notation, we are not dealing with works. Tartini's Sonata is music, but not a musical work. As such it cannot be used as a counter-example to challenge a theory specifically of works. Were Goodman to claim this, and were he justified in claiming this, then, like the first solution involving translation, this second one would undermine Ziff's challenge.

Goodman proposes the second solution less for early music than for much twentieth-century, avant-garde music. Many critics have highlighted various features exemplified in avant-garde (specifically aleatoric, randomly generated) music to reveal the large discrepancy between Goodman's theory and musical practice. Goodman himself spends time considering the notationality of this music (187–90). His conclusion is that the language in which certain scores of avant-garde music are written does not satisfy his notational requirements and therefore is not notational in his terms. Sometimes, he writes, 'in some of [John] Cage's music that uses non-notational sketches in place of scores, we have no . . . definitive criterion and so in effect no works'.[31] Such music fails the retrievability test.

Following along with Goodman, it is no longer clear that all or any pieces of avant-garde music count as bona fide examples of works. Thus, it is not worth considering in a blanket fashion whether avant-garde scores are notational or not if the purpose of the consideration is to judge the adequacy of a theory of *works*. If much avant-garde music is not packaged in terms of works, the relevant items cannot provide for Goodman's critics counter-examples to his theory. By extension the same argument applies to any music performed without the use of 'adequate' scoring. If Tartini's Sonata fails the retrievability test, it does not function as a work.

Goodman's critics are dissatisfied. They are being asked to accept that any example of what we in practice call a work, produced to serve as a counter-example, can always be rejected by Goodman as an illegitimate example. The critics cannot win. Is there any way, then, to challenge Goodman on what he would see to be legitimate grounds? Perhaps one could show that his theoretical use of terms is in itself problematic, or that one has a more satisfactory account.

XII

I suggested above that 'inadequate' scores of early music could be translated into 'adequate' scores. Let us develop this solution, but with regard now to aleatoric music. Both kinds of music—early and aleatoric—share the property of inadequate scoring, in Goodman's sense, at least for some cases.

Not all scores in modern music notation are inadequate. Many scores and increasingly computer programmes play a central role in determining the identity of aleatoric music. Carlos Chavez described in 1937 how with electronic equipment, one could create music of fixed values that would be unalterable in successive performances.[32]

---

[31] From Goodman's 'Comments on Wollheim's Paper', in R. Wollheim, D. Wiggins, and N. Goodman, 'Are the Criteria of Identity that Hold for a Work of Art in the Different Arts Aesthetically Relevant?', *Ratio*, 19–20 (1978), 49. On pp. 180 ff. of *Languages of Art*, Goodman draws a parallel between avant-garde and early music notation.

[32] 'Mechanical Performance and Fixed Music', *Toward a New Music: Music and Electricity*, tr. H. Weinstock (New York, 1975), 60 ff.

We have a situation here in which the same electronic tape is played over and over again in each successive performance. And we can imagine the case of a computer programme where the output is identical in its constitutive attributes in successive performances. These kinds of programme or algorithm have been produced in recent years, and they serve to reinforce the emphasis on notation albeit now somewhat more broadly viewed. This view of notation supports Goodman's theory.[33] It even increases the means by which one could translate inadequate scores into adequate ones. We could simply produce for all music computerized scores containing neither ambiguity nor imprecision.

It would not even be a problem that avant-garde notations differ from traditional ones. Often constitutive characters are not tonal but rather rhythmic, dynamic, or durational. Many avant-gardists have tried to move away from the idea that two performances of a work should 'sound the same'. But from the fact that two performances do not sound the same it does not follow that they are excluded from being, in other respects, isomorphic or replicas of one another. Isomorphism may obtain in many respects—remember there can be more than one musical language.

We could take a step further and suggest, in the light of the changes in notation shown in the work of Cage, Henze, and Stockhausen, that scores be understood in terms of rules rather than characters. For example: 'Tap bow on back of violin for 20 seconds.' Rules suitably conceived might allow for a certain kind of specification not captured by traditional characters. A musical work, we could now say, is a class of performances in which all the appropriate rules are followed. Whether or not the rules are formulated in the traditional manner in terms of timbre, pitch, and key would now be an *option* for the composer. It would not be *condition* for producing musical works as such.

XIII

Alan Tormey has provided compelling reasons for replacing the character with a rule model of the score. He does so, however, not to support but rather to offer an alternative model to Goodman's. First, he rejects Goodman's solution that inadequate scoring (in Goodman's

[33] See, however, his own comments regarding the problem of semantic density (190).

terms) renders the music something other than a work. It does not follow from the fact that aleatoric scores fail to meet Goodman's notational requirements that 'the system furnishes no means of identifying the work from performance to performance'. We do identify aleatoric *works*. So 'there must be alternatives to a notational warranty for identification'. Second, Tormey shows why a rule model would account not just for the difficult cases that do not fit Goodman's model but for all cases. The rule model has, he argues, one special advantage over the character model. It reveals all those crucial *non-notational* 'safeguards of identity'.[34]

Tormey makes a critical point against Goodman. We do not need to translate inadequate scores into adequate ones, because there are conditions other than notational accuracy that help to preserve the identity of works. We can include more kinds of music in the class of works if we demonstrate that notation conceived along Goodmanian lines is insufficient to explain the identity of a work in a chain of performances. In addition to notation we require non-extensional conditions having to do with beliefs and intentions. Admitting these renders the extensional, notational conditions more flexible. Non-extensional conditions can help determine the identity of a work or a performance in cases, say, of overlapping classes.

A great deal is packed into the word 'identity'. Like Goodman, Tormey is not speaking of identifying a work or a performance according to what is heard, what we read in the programme notes, or what we are told we are going to hear. He is using the term technically; something has the identity it has if and only if it meets certain extensional or non-extensional conditions, or both. The difference between Tormey and Goodman turns upon what sort of conditions they are willing to count as relevant to a work's identity.

Would Goodman's claim, that the score is the authoritative principle of identification of a work, be affected were compliance with characters to be replaced with satisfaction of rules? Perhaps it would threaten his extensionalist commitments. But if he could articulate a rule model in extensional terms, rules might end up doing just as well as characters. Probably he would argue that no genuine distinction exists between what he calls characters and what others refer to as rules. So where's the rub?

---

[34] 'Indeterminacy and Identity in Art', 203–15.

Perhaps here. Goodman's notational system allows for a two-way identification of a work through the retrievability process. One can retrieve performances on the basis of reading score-copies, and score-copies on hearing performances. Would replacement of a character by a rule model threaten this two-way identifiction? On hearing a performance could one identify which rules had been followed? I doubt it. But if this is a problem for the rule model, it is also a problem for the notational model. Ziff once criticized Goodman's theory on the grounds that one cannot extract a work simply on the basis of hearing a performance, by reference only to the notational properties. One has to look in addition to the tradition in which the work was created. Thus he concluded, as Tormey does, that the score cannot by itself define a work.[35]

Goodman would grant that in reading scores or hearing performances one has to have an understanding of what counts as the constitutive properties of the work. Further, one does not simply read black marks on paper but meaningful signs (117). One has to presuppose that we know the rules for transforming written marks into musical sounds, that we know what is involved in producing, say, a compliant of Middle C. It is to be presupposed, in other words, that the meaning of the notational characters is assessed within the context of the relevant tradition. Similarly, on hearing performances, it is to be assumed that we (or at least some persons) know which characters the sounds comply with.

Whether Goodman would actually grant all of this has little bearing on his theory, however, since the retrievability test tells us only what theoretically must be possible. If the score is to serve its function, it must in theory be possible to retrieve a score from a performance and vice versa. Nothing is said about what has to be possible in practice.[36] One could, only for interest's but not for theory's sake, recall the case of Mozart, who, on hearing a performance of the *Miserere* by Gregorio Allegri, returned home having memorized the piece and was able (at the risk of offending the Pope, who had forbidden its public scoring) to produce a score-copy of it. On being presented to the Pope, the score-copy was judged to be note-perfect.[37]

[35] 'Goodman's *Languages of Art*', 509–15.

[36] Cf. the kind of sentence to be found on p. 154: 'But none of this has anything to do with the basic theoretical function of notational systems.'

[37] For further details of the Mozart story, see E. W. Galkin, *A History of Orchestral Conducting: In Theory and Practice* (New York, 1988), 457. Peter Kivy uses the story to make a theoretical point in his 'Platonism in Music', and I believe others do so as well.

But then proponents of the rule model could similarly defend themselves. The rule model requires only in principle that the retrievability test be passed. It is part of understanding a rule that we recognize, in principle, when a rule is being followed and are able to identify that rule. Whether we can correctly identify the rule in practice is another matter. Hence, if the retrievability test serves only to make a point about theoretical possibilities, it will not help us decide between the character and rule models, for both seem to pass it equally well—in theory.

Were we to claim no prior allegiance to extensionalism, we could investigate the possibility of non-extensional conditions (in addition to extensional ones) aiding the retrievability procedure. The debate would no longer be about how to judge between two models conceived in the same extensional terms, but how to judge between two models bearing perhaps quite different ontological commitments.

From a different angle, Tormey's rule model has the advantage of accounting for aleatoric and early music scores, and that already counts for quite a lot. Goodman, however, would not see this as a genuine advantage. We have shown that he has ways to deal with difficult examples—through translation or banishment from the relevant category. Apart from these solutions there is his other reason why he is not worried by mention of examples unaccommodated within his theory. His intention has been to produce an account of the identity of works based on an idealization of the world of music. He has not been directly concerned with what is counted as a work in the real world, nor with whether there are any works in the real world meeting his theoretical conditions.

Goodman's critics still cannot accept this. His idealization of the musical world fails to retain the kind of theoretical purity he desires it to have. Idealizations are, after all, idealizations derived from our knowledge of real world examples. Despite Goodman's own proviso that he is not 'primarily concerned with the origins or development [of musical notation] but with how fully the language of musical scores qualifies as a truly notational system' (181), many of his claims still presuppose knowledge of real examples. Were Goodman to deny that he presupposes any empirical knowledge at all, and were he to affirm that his use of the terms 'work', 'score', and 'performance' is legitimate only within theory, we could ask him what he has been talking *about*. What is his theory a theory of ? Do his claims have anything to do with actual musical works? If they do, wherein resides the connection?

If they do not, why does he use these particular terms? Is his use not apt to confuse those who associate non-theoretical meanings with the same terms?

Goodman would undoubtedly remind his critics now that there are certain demands a theory about any worldly phenomena must meet which can sometimes override one's loyalty to the phenomena themselves. Often he says this much.[38] But to understand fully the background and the ramifications of this rejoinder we need to turn to his account of musical performances.

XIV

Goodman's concerns are not with how we—as potential performers—bring about good or even any performances, just as his concerns are not with how musicians read scores. Just as he was concerned to specify the conditions something must meet to function as a score, so now he wishes to specify the conditions something must meet to be a performance. Accordingly, he argues, the relation between a notation and its compliants must be complete or perfect.

The competence required to identify or produce sounds called for by a score increases with the complexity of the composition, but there is nevertheless a theoretically decisive test for compliance; and a performance, whatever its interpretative fidelity and independent merit, has or has not all the constitutive properties of a given work, and is or is not strictly a performance of that work, according as it does or does not pass this test. (117–8)

And later, he writes:

Complete compliance with the score is the only requirement for a genuine instance of a work . . . Thus while a score may leave unspecified many features of a performance, and allow for considerable variation in others within certain prescribed limits, full compliance with the specifications given is categorically required. (186–7)

For Goodman there are properties of a work for which the composer allows no 'free play'. Typically these are tonal or rhythmic properties and properties associated with key. These properties need to be

---

[38] Goodman is content, in places, to refer to 'standard musical notation' (215) or 'the standard language of musical scores' (186). But he also writes: 'As with most familiar words, systematic use involves a specialization from ordinary usage' (n. 128). He then suggests that examples conforming to the ordinary usage of terms have, in a systematic account, to be reclassified.

specified precisely since they count as the work's constitutive properties. They have to be exhibited in every performance. Those properties for which a certain degree of variation, interpretative freedom, or *'laissez-faire'* is admissible are the contingent, unforeseeable ones (190). The fact that in their scores and performances composers allow a certain openness is not thought by Goodman to threaten the notational character of the musical system. As long as the notational requirements are met by the characters in the scores the authoritative identification of works can in principle take place. Is there a problem here?

XV

Imagine a sound event taking place in a concert hall. During the event one of the orchestra players sneezes so that the total sound event produced consists of sounds compliant with the score plus a certain noise. Does this sneeze serve to violate the condition of perfect compliance, rendering the event something other than a performance of the work?[39]

The question has a negative answer if we allow that the score prescribes the constitutive properties of a performance, and that something counts as a performance if and only if it complies perfectly with these properties. That every performance also has contingent features favourably implies that two performances will rarely if ever sound exactly the same. It also implies that any feature can be added to the sound event if it does not get in the way of the constitutive features being complied with; the consequent event is still a performance. This is an undesirable conclusion. Or is it?

Goodman would probably say this conclusion is undesirable only in pragmatic or qualitative terms. He would probably remind us that a score is a compound character and that when we perform a work it is less a matter of complying with each individual character than of complying with characters in their modes of combination. If performers comply with the constitutive characters, they have complied with them in their modes of combination, and that might be sufficient to

---

[39] What this would be called is problematic—a non-performance? an incorrect performance? Goodman calls it an incorrect performance; he explicitly provides conditions for correctness of performance (186–7). The distinction between correct and incorrect performances is used by many theorists. A correct performance is also contrasted with a good performance, where the former has to do with identity and the latter with aesthetic merit. See Walton, 'The Presentation and Portrayal of Sound Patterns', 241 and Wolterstorff, *Works and Worlds of Art*, 59 and 64.

rule out the possibility of there having been a sneeze, say, between every other note.

Perhaps an isolated sneeze can appear as one of those contingent features that, in Goodman's terms, affects the performance's quality but not its identity. Where violinists choose to add a little extra vibrato, so they might choose also to add a sneeze. Perhaps, furthermore, a distinction can be introduced to prevent this choice, one, say, between properties that are acceptably *notationally contingent* in the sense of undetermined (they are the properties for which a certain freedom of interpretation is allowed) and *accidental* properties that bear little or no relation to notational specifications at all, such as sneezes, coughs, thunderclaps, squeaking doors, and paper rustles.[40] A performer is free to use independent judgement with regard to notationally contingent features; accidents are another matter. The latter are treated as disruptive and irrelevant. Does the distinction help? Does Goodman even need it? To both questions the answer is no.

In Goodman's account, one is obliged to account for performances only in terms of the constitutive sounds produced and the degree of their compliance with the relevant scores. Which contingent sounds are produced is irrelevant. If a sneeze or an overdone glissando enters the audible space in which a performance is taking place, they violate the performance's identity only if they interfere with compliance with constitutive characters. When they do interfere, as they inevitably do on occasion, this is a practical consideration and we judge the performances accordingly—as incorrect or imperfect. But that there are incorrect performances in the real world has no theoretical bearing on whether perfect compliance suffices for, and determines, performance identity.[41]

Goodman's account cannot, moreover, be undermined because of epistemological difficulties. One cannot produce examples where we are not able to judge whether or not it is a performance, because

---

[40] Such extra sounds could also be musical sounds. In principle, an orchestra could produce a performance of a Beethoven symphony even if, between each movement, they played the just-played movement backwards. This would be possible if the (sneeze) interval the orchestra took between each movement was neither accurately specified nor limited by some other prescription.

[41] At Tanglewood Music Festival 1987, a triumphant performance of a new fanfare by George Perle was interrupted by the noise of a motorcycle. Though the audience recognized that a performance· had in fact taken place, another performance was thought appropriate and duly given.

Goodman will retort that this consideration is irrelevant. How we *know* whether a sound event complies with a score has no bearing on the ontological condition which demands that performances comply perfectly with such scores. By extension the point applies to the distinction between notationally contingent and accidental properties. Whether one chooses to exemplify either kind of property makes no difference to the identity of a performance, only to its quality.

It turns out that the sneeze is not a problem for Goodman's theory so long as its author ignores all pragmatic considerations. Critics remain dissatisfied. More important than the sneeze, they want to count as performances many events that do not perfectly comply with the relevant scores. This turns out to be the core of their disagreement with Goodman. Goodman allows any number of sneezes (or additional sounds) so long as they do not interfere with compliance, but he refuses to allow a performance with a single mistake (even a single wrong note) to count as a performance. What theoretical issue demands that he take this position?

XVI

A famously problematic implication of meeting the perfect compliance condition is that any performance, however boring, satisfies the notational prescription so long as it has no mistakes. Contrarily, the most brilliant performance, if it has but one mistake, does not count as a performance of the work. This has to be the case, Goodman argues, if a work's identity is to be preserved.

The innocent-seeming principle that performances differing by just one note are instances of the same work risks the consequence—in view of the transitivity of identity—that all performances whatever are of the same work. If we allow the least deviation, all assurance of work-preservation and score-preservation is lost; for by a series of one note errors of omission, addition . . . we can go all the way from Beethoven's *Fifth Symphony* to *Three Blind Mice* (186–7).

A score must specify at least everything that is constitutive of a work, and a performance must comply perfectly with each and every one of the score's notational specifications. It is an implication of the familiar sorites (or heap) problem that there is no legitimate point at which it would be admissible to assert that a performance fails to comply with one of these specifications. By what criterion could one decide, by reference to the score alone, how many mistakes were admissible in a performance of a work?

Consider the case of a person with a full head of hair. How many hairs may fall out before we would say of that person that he no longer has a head of hair, that he is bald? For any answer given one can always ask why not one more or one less. Similarly one can ask how many wrong notes may be played before one would say of a sound event that it was not a performance of the work. The principle that two things belonging to a given class are individuated in the same way even if they differ minimally must be rejected. Otherwise, given that class membership is transitive, one would logically have to conclude that a man with a full head of hair is bald, or that Beethoven's Fifth Symphony is identical to Three Blind Mice.

One difference between a head of hair and a performance (apart from the fact that there is no hirsute–bald equivalent in music) is that all the parts of the latter are interrelated. Perhaps in performances, or in the works themselves, it is not the individual tones that are the primary units but, rather, the intervallic structures—the rhythmic or harmonic *Gestalten*—even the melodies. If the identities of the more complex parts are retained, we can admit mistakes in individual notes. So long as we recognize the melody-*Gestalt* it does not matter that one or two notes are incorrect.

This proposal has problems. First, recognizing *Gestalten* usually does not fit extensionalist description, so let us give up this notion and speak just of compliance with complex units. Second, the sorites problem is raised anew at the higher level: if we miss out this melody, or this thematic structure, will the performance's identity be retained? Even if we could answer this question we would soon have to explain how a melody's identity is explained other than by reference to its parts, albeit in their interrelation. To instantiate a melody must we not instantiate each atomic part?

Given a quantitative model, according to which we count the number of wrong notes, perfect compliance might after all be the most satisfactory condition. Is this the correct conclusion to draw? Would Goodman's theory be affected were we to stipulate that compliance be at least, say, 80 per cent?[42] Could we individuate between performances each as being of a given work if we allowed this degree of flexibility or margin of error? Why we would choose this particular percentage as

[42] I have benefited here from D. Parfit's discussion of personal identity in his *Reasons and Persons* (Oxford, 1984), pt. III.

opposed to any other is not obvious, however. And this seemingly arbitrary choice might be theoretically less satisfactory. After all, there might be hidden in Goodman's theory another reason for adopting the perfect compliance condition other than that bearing on the sorites problem.

XVII

Goodman might be offering the condition of perfect compliance to avoid a commitment to vague objects. As Nathan Salmon writes: 'Sorities [sic] arguments are notorious for playing havoc with the phenomenon of vagueness.'[43] Intuitively we want to allow that the concept of a performance-of-a-work be such that objects falling under it need not share an exact community of (constitutive) properties, that something else explains its use in each given case. If performances of a given work do not end up exhibiting exactly the same (constitutive) properties, they none the less are identified as being performances of the same work in virtue, perhaps, of the work which they *intend* to instantiate and which we *recognize* them as instantiating. Someone inclined towards platonism could argue that it is the work that has a fixed, unalterable set of properties; the performances are only the imperfect approximations or copies. We think of musical works, writes Wolterstorff, as 'immutable with respect to the properties essential within them, for they cannot be different in this respect from how they are'.[44]

This argument has pre-critical intuition on its side. Works are thought to be the exactly specified products of composers. A composer's decision regarding the specification of a work's structural (and as far as possible its aesthetic) properties is generally respected. Still, performances are not expected to be perfect (even though performers strive to make them so), because their imperfections can be judged and even ignored given the perfect, determinate, and idealized conception listeners have of the work itself.

Goodman cannot recognize intentional and recognitional conditions and straightforwardly preserve his extensionalist commitments. Nor can he simultaneously countenance the openness or vagueness of performances and the rigidity or perfection of the work, because the

---

[43] *Reference and Essence* (Oxford, 1982), 240. Goodman makes no explicit reference to this problem in his *Languages of Art*.
[44] *Works and Worlds of Art*, 88–9.

latter has ontologically been reduced by him to the former; a work is just those performances compliant with a score. Goodman has then a difficult choice between two options, both of which conflict with our intuitions. He can allow either that there is a degree of vagueness in both the work *and* its performances, or that they are both equally and perfectly determinate as regards their constitutive properties. He opts for the latter.

Goodman is thus obliged to relinquish loyalty to our pre-critical understanding of musical practice. He does not mind this. He writes:

The composer or musician is likely to protest indignantly at refusal to accept a performance with a few wrong notes as an instance of a work; and he surely has ordinary usage on his side. But ordinary usage here points the way to disaster for theory. (n. 120)

Goodman's critics remain as ever dissatisfied. They accept neither his loyalty to nominalism and extensionalism, nor his willingness to break completely with one's pre-critical intuitions about musical phenomena. The discrepancy between his theory and how we think about musical practice is just too great. Their dissatisfaction places the onus upon themselves, however. Do they have an account that is theoretically coherent and maintains a satisfactory relationship to musical phenomena? It is this question that has motivated the production of numerous theories in the last two decades. In the next chapter we shall consider one of the most plausible theories to have been offered thus far.

2

A Platonist Theory of Musical Works

Jerrold Levinson offers a modified Platonist view. He argues that musical works exist as abstract entities apart from their performances and scores, and that the identity and individuation of performances involves intentional and recognitional components. Both claims pave the way for an ontological account unavailable to a nominalist and extensionalist like Goodman. I shall treat Levinson's account of works and performances separately and I shall pay more attention, as the author does himself, to the former than the latter. As in the previous chapter the purpose is not to seek the best analytic theory of musical works. It is to discover what analysts think are the important questions and most satisfactory kinds of answer when they produce theories of this sort.[1]

I

Levinson argues that musical works, once created, are fully formed and permanently existing entities. They are unchanging continuants. Wolterstorff also claims that we think of works as capable of existing unperformed, and, he adds, we do not think of them as existing intermittently. Why do we want works to be thought of in this way? Neither theorist provides a detailed answer to this question, probably because they feel the view harmonizes so naturally with our pre-critical intuitions that it is in need of no further defence. Levinson notes only that

works are the main items, the centre and aim of the whole enterprise, and that since musical works are not identical with scores, performances, or thoughts, if those are the only things actually created, then much is lost (9).[2]

---

[1] For the rest of this chapter, unless otherwise indicated, all parenthetical page references will be to Levinson's article 'What a Musical Work is'.

[2] Levinson does not specify further, but he is obviously referring to changes in evaluation that would result if works were not considered separately from their performances. Walton concurs with this reading: 'To call the Bartók String Quartet delightful . . . is to describe the Bartók String Quartet itself, not merely to generalize about its performances' ('The Presentation and Portrayal of Sound Patterns', 245). Cf. Ingarden, The Work of Music, 5–6, for similar discussion.

Levinson rejects the idea that performances or scores are the only things that exist. A work is a distinct entity, something to which we can attribute properties, properties attributable to nothing but the work. He does not want to see the composer as being the original generator just of a chain of performances and score-copies; the composer produces some thing, a unique product, in addition. Central to his theory, then, is a basic conception of works as distinctly existing objects.

Can Levinson support this conception in ontological terms? He believes so and accordingly adopts the following strategy. He bases his account on two kinds of argument. Some claims are supported by what he takes to be non-negotiable intuitions. These intuitions are derived from our pre-critical, practical, or aesthetic understanding of musical works. Where the intuitive arguments are weak or muddy they are strengthened with theoretical arguments. In this way Levinson intends to retain the balance between his pre-critical intuitions about musical works and their ontological explication. To what extent does he succeed?

II

For many who have sought to provide an account of the ontological status of musical works there has been a temptation to identify the works simply with the patterns or structures of sound indicated in scores and played out in performances.[3] Levinson does not yield to this temptation. He argues for the separate (though related) identities of works and sound structures. His argument is based in its initial stages on our accepting three basic claims. Works, but not sound structures,

  (i) do not exist prior to the composer's compositional activity, but are brought into existence by that activity (9);

 (ii) are such that composers in different musico-historical contexts, who determine identical sound structures, invariably compose distinct musical works (14);

(iii) are such that specific means of performance or sound production (instrumentation) are integral to them (19).

With regard to (i): that the compositional factor plays a role in the determination of a work's identity is a claim typically supported by

---

[3] Cf. Levinson's remarks on claims made by Walton and Wolterstorff (6 n. 4).

those theorists who require that some historical facts about the origin of a work are relevant to the determination of its identity', theorists influenced perhaps by so-called essentiality of origin theses.[4] Levinson by contrast argues on intuitive grounds. Much of the 'status, significance and value' we attach to musical works is bound up with our belief that works are necessarily the products of a particular person's compositional activity. This belief is 'one of the most firmly entrenched of our beliefs concerning art' (8).

When we appreciate a work our appreciation is affected by knowledge of who composed it. We do not appreciate a work as existing atemporally, but as created. Levinson stresses the importance of the compositional factor by inverting for rhetorical purposes the infamous 'argument from design'. 'Art is creative in the strict sense,' he argues.

[It is] a God-like activity in which the artist brings into being what did not exist beforehand much as a demiurge forms a world out of inchoate matter . . . There is a special glow that envelops composers, as well as other artists, because we think of them as true creators (8–9).

Levinson is making not only a rhetorical but also an ontological point, in fact, two ontological points: first, works have as a generic *identity* condition that they each be created; second, that each work is *individuated* (at least in part) according to the particular composer who brought it into existence. Each work is 'necessarily personalized' (28).[5]

With regard to (ii): much of the value we attribute to a work has its source not only in the fact that a particular composer created it but also, Levinson argues, in the fact that the composer is representative of a particular time and place. Works are composed against a background of particular historical epochs and are thought to be expressive of these. Musical works are characterized with reference both to their directly exhibited, auditory properties that are captured by a score, and to their non-exhibited, relationally based properties, such as 'being classical', 'being revolutionary', and 'being Liszt-influenced'. The latter as much as the former determine the aesthetic character of the work (12).

---

[4] See Lord, 'A Kripkean Approach to the Identity of a Work of Art'.
[5] Cf. J. C. Anderson, 'Musical Identity', *Journal of Aesthetics and Art Criticism*, 40 (1982), 285–91. Anderson considers Levinson's argument in the light of 'possible worlds' or 'cross-world identity' semantics.

These non-exhibited properties are partly determined by a work's placement in a particular historical or musical tradition. This determination is based on our conceptions of style, œuvre, and genre each of which is socially and personally inflected. The judgement that a work is original depends for its sense on its being produced at a given time, its originality assessed in relation to the history of music preceding it. The idea of a living tradition is crucial to our understanding, experience, and evaluation of musical works. Yet non-exhibited properties, and related judgements making reference to a musical tradition, cannot be spoken of with regard simply to the sound structure of a given work. Thus, Levinson reminds us, the work and its constitutive sound structure are not identical.

With regard to (iii): Levinson is led to conceive of a musical work as ontologically related to a combination of sound and performing-means structures. Integral to a work are its sounds and rhythms and its instrumental colourings. We do not think, Levinson claims, that instrumental specifications are arbitrarily chosen by composers. These specifications, as much as the specifications of sound, help determine the aesthetic character of the work. 'The idea', Levinson suggests, 'that composers of the last 300 years were generally engaged in composing pure sound patterns, to which they were usually kind enough to append suggestions as to how they might be realized, is highly implausible' (15).[6] Later, he adds, 'if we want . . . pieces to have the definite aesthetic qualities we take them to have, instrumentation must be considered inseparable from them' (27). If, moreover, the performing-means structure is a constitutive part of the work, the work cannot be exhaustively identified with its sound structure.

III

With the three desiderata clarified, Levinson moves on to describe the ontological status of works. Though related to both, a work is straightforwardly identical neither to the sound structure nor to the performing-means structure, nor even to a combination of the two. A

---

[6] 300 years is an exaggeration. Probably Levinson would agree. At one point he remarks that from 1750 on, compositions involved 'quite definite means of performance', and that Bach's *Art of the Fugue* is 'merely the exception that proved the rule' (15 n. 18). 1750 to the present is nearer to 200 years. This is more accurate, and in this book it is an important difference. In a discussion of the same topic, Wolterstorff speaks of 200 years (*Works and Worlds of Art*, 69–70).

work is rather *derived* from or *constituted* by these structures. What is the difference between the 'is' of identity and the 'is' of derivation?

Levinson suggests that both sound and performing-means structures are *implicit* or *pure* types. Their existence is implicitly granted when a general framework of possibilities is given. Given that individual sounds exist, the existence of all possible combinations of these sounds is implicitly granted; given that instruments exist, all combinations are implicitly granted; given, finally, the existence of sounds and instruments, all possible combinations of sound and instrumental properties are implicitly granted (21). To describe implicit types in this way is to imply that a combination, be it of sounds or instruments or both, is no more than the sum of its parts. It is also to make a logical point. Levinson is not saying these combinations concretely exist in the world of tables and chairs, though they might be instantiated in that world. He is saying that they exist as potentialities or possibilities of concrete combination. They exist as something like unrealized forms, that might in fact be realized.

Implicit types do not exhibit properties that we find in the realm of existing combinations. Such properties would include, for example, those indicating a type's moment of 'being noticed, recognized, mentioned, or singled out' (22). Rather, Levinson argues, the sound and the performing-means structures are 'types of a pure sort which exist at all times' (7–8). He continues:

This is apparent from the fact that they—and the individual component sound types that they comprise—can always have had instances . . . If the pre-existence [before, say, its realization in a musical work] of simple sonic element types be granted—and I think it must be—it follows automatically that all sets and all sequences of these elements also pre-exist (7).

This characterization of implicit types does not tally with Levinson's desiderata for the role played by creativity and historical location in the determination of a work's identity. Thus, he argues, musical works are not implicit types. Rather they are *initiated* types *derived from* implicit types.[7]

---

[7] Levinson also says that initiated types *realize* implicit types. Levinson's description is similar here to Wolterstorff's, yet they differ regarding the ramifications their descriptions have for a work's ontological status. See what I call Case 2 below. Levinson also characterizes the distinction between implicit and non-implicit types in terms of a distinction between structures and constructions. 'The Brooklyn Bridge did not exist before its construction. But the geometrical structure it embodies, which required and received no construction, has always existed' (8 n. 9).

Levinson distinguishes implicit types, whose existence is taken to pre-date or to exist prior to any specific creative activity, from those types whose identity is determined by factors having to do with the manner and time of their initiation. The initiated types might come into existence via a compositional act or an act of making. All those *initiated types* can 'be construed as arising from an operation, like indication performed upon a pure structure [an implicit type] . . . Poems, plays and novels—each of these is an entity more individual and temporally bound than the pure verbal structure embodied in it' (21). Levinson opts ultimately for the description of a musical work as a 'sound/performing-means structure as indicated by X at t, where X is a particular person—the composer—and t, the time of composition' (20).

IV

Levinson's account gives rise to many questions, some of which can be articulated in the form of objections. In saying, for example, that musical work types are derived from implicit sound and performing-means structures, Levinson commits himself to the idea that the combinatorial relations between notes are nothing more than mereological relations. But, one common objection goes, the relations between musical sounds are at least as important if not more important than the relata. Retaining, say, the intervallic relations between notes is often more important than retaining the specified pitches. Further, the relational elements in a musical system do not exist just as a logical, summing consequence but are as much historically and culturally conditioned as the works themselves.[8]

Levinson would probably argue that there is nothing in his account that precludes his speaking of relational elements of a musical language. On the contrary: works derive from the combination of many implicit types—pitch, duration, rhythm, interval, and timbre types.[9] This would be a satisfactory response, though on its own it does not meet the whole objection, the main import of which was that intervallic structures, and musical languages more generally, do not exist eternally or ahistorically, but within living and changing musical practices.

[8] This sort of objection, and a related discussion of closed (finite) and open musical systems, can be found in R. Cox's 'Are Musical Works Discovered?', *Journal of Aesthetics and Art Criticism*, 43 (1985), 367–74.

[9] 'It should be understood . . . sound structure includes not only pitches and rhythms, but also timbres, dynamics, accents' (6 n. 3).

Levinson might now remind his critic that his use of the notion of implicit existence is *required* on logical grounds and cannot therefore simply be adapted to account for the historical character of a given musical language. Certainly he uses the principle that 'given that all *X*s exist—given that framework of possibilities—then the existence of all combinations of *X*s is implicitly granted'. But all this means is that if a composer uses a musical language, say, of twelve tones, and the use of this language is not restricted by rules preventing certain combinations of these tones, the composer is free to make use of any and all combinations of them.

The critic might still force Levinson to acknowledge and then account for the fact that musical languages are themselves historical. To make a logical point is one thing, to account for the historical character of real musical systems is another. Saying that if a twelve-tone language exists, the existence of all combinations of these twelve tones is implicitly granted, strips the language of its artificial, conventional, or historical character. Besides, if one accounts for the historical character of the individual works, why not do the same for the languages in which the works are expressed?

Levinson might retort that were he to provide a more detailed account of the composition of works, he would speak of both the initiation of work types and that of language types. There can be and indeed are many different kinds of initiated types of a more general or specific nature. Many works, though unique in their particular constitutive structures, nevertheless use common musical languages, and these, as much as individual works, are human constructions. Musical languages—systems of atomic and compound elements—are *not* pure or implicit types that have always existed in a logical sense. They only derive from such.

In this additional view, the structures constitutive of given musical works are derived *indirectly* from pure, implicit types and *directly* from those initiated types constituting an existing musical language. Initiated languages mediate between implicit types and individual musical works. Works are expressed directly in a musical language. The language itself is an initiated type derived from an implicit musical language type (or a class of implicit sound, timbre, and dynamic types). Obviously language types are not *identical* to work types. The latter exhibit a range of historical and aesthetic properties not attributable to the language within which they are expressed. Though initiation might occur at the same time in the same place, the manner of their respective initiation is different. Constructing a musical language is not the same activity as constructing a work, even if on occasion by constructing the latter one constructs the former.

One advantage of extending Levinson's account in this way is that it shows how works acquire a full range of aesthetic properties. A given work has a personal character formed as a result of an individual's compositional action and a social character as well, resulting from a composer's use of a public musical language. That works have different kinds of aesthetic properties is an indispensable fact when we want to assess such things as individual and shared styles. We refer to different sorts of properties when we assess the style of a single work (if individual works can be said each to have a style), the style of a composer's output taken as a whole, or the style of a genre or a period.[10]

Apart from its intrinsic interest, the entire proposal for the extended account is designed to show that Levinson can adapt his theory to meet the objection regarding the historicity of musical languages without compromising his ontological position, specifically the distinction between implicit and initiated types. Unfortunately there are other more profound objections to his account that cannot so easily be dealt with.

V

Levinson places musical works in an ontological category of initiated types that is neither straightforwardly universal nor particular, as these categories have traditionally been conceived. His musical works have all the ontological (formal) properties that make them of universal status but, because of their cultural skin and aesthetic character, they are pulled out of the ahistorical and ahuman world of eternally existing types and brought into a still abstract but now historical world.

Theorists might, for metaphysical reasons, resist the positing of a new category even if, for reasons derived from aesthetic concerns, they do not. Three questions arise: is Levinson constructing an ontological category just to match his description of musical works, or does he have independent reasons for so contructing it? Does he treat aesthetic issues satisfactorily whether or not they bear on ontological

---

[10] Cf. Walton, 'Style and the Products and Processes of Art', in B. Lang (ed.), *The Concept of Style* (Philadelphia, 1979), 45–66.

issues? To what extent is he justified in taking aesthetic considerations seriously while he deals with issues of identity? I take these questions in turn.

At some point Levinson must confront the age-old problem regarding the source of ontological knowledge. Are ontological categories posited on a priori, transcendental, or logical grounds where grounds of these sorts are contrasted with those derived from empirical observation of the relevant phenomena? On the former grounds one might work out all the various ways phenomena could exist given a range of acceptable conditions and then fit the categories to the phenomena. These conditions would be derived from answers given to questions concerning ontological commitment, spatio-temporal existence, and the nature of concreta and abstracta. On the other grounds one would begin with the phenomena themselves, or at least pre-ontological descriptions of such, and then see how many ontological categories were required in order to account for them all.

Within the confines of an article Levinson could not be expected to address the general issue concerning the source of ontological knowledge directly. Neither should one have expected him to opt straightforwardly for either crudely described route of inquiry. Still, he does seem content to posit a category of object unfamiliar in traditional metaphysical frameworks to account for what he sees in pre-ontological terms to be essential to the phenomena in question, and this is sufficient to justify our concern. Is this extra category required? Has Levinson accounted for musical works in more adequate ontological terms than Goodman? The last question seems to be a bona fide one. On closer inspection it turns out to depend upon our accepting a quagmire of unclarified assumptions, making its answer impossible to give. We are not yet at a point where this quagmire can be fully displayed. We need first to locate some pointers to help us on our way.

Levinson appeals to the original context of production—both individually and socially inflected—to partially account for a work's identity. It helps him to explain, furthermore, how a work acquires a determinate and fixed aesthetic character. This explanation turns out to be entirely consistent with his traditional essentialist or objectivist view of the meaning of music and art. It is incompatible, however, with the conflicting view, namely, that a work is a potential source of meaning lacking in itself a determinate aesthetic meaning and character.[11] That

---

[11] Levinson adds an important caveat (25 n. 29): 'aesthetic attributes do not have to be regarded as *essential* to works, but they do have to be regarded as *relevant* to

Levinson assumes a connection between his three desiderata and a particular aesthetic theory is telling. For what we find is that Levinson's initial pre-critical claims about musical works are pre-critical relative only to *ontological* theory, but not to *aesthetic* theory. Yet if ontological and aesthetic considerations are to be combined, if ontological claims are to be based even partially on aesthetic claims, eventually the latter will have to be argued for to render the theory fully critical. Levinson himself remarks that something like the principle that '[w]orks of art *truly have* those attributes that they appear to have when correctly perceived or regarded' (11 n. 15) serves well enough to justify his general view and ultimately the use of his three desiderata. To this he adds the disclaimer that this principle has been sufficiently well argued for by other theorists. Despite this disclaimer it remains unclear, however, what something's *truly* having attributes really means.

Even were Levinson to provide the requisite justification for his aesthetic claims, this would not automatically serve as a justification for his using these claims as anything more than guidelines in his ontological inquiry. Are not questions about aesthetic character independent of those about identity? With what justification can Levinson posit an ontological category on the basis of a work's having particular aesthetic attributes?

Levinson could distinguish two points: first, a work is identified in part according to the musico-historical context in which it was composed; second, a work has a particular aesthetic property, say,

individuating between them—where this means that (in at least this world but not necessarily in all possible worlds) aesthetic and artistic attributes *truly belong* to works in a *reasonably determinate* fashion.' One might want to look at Levinson's recent article 'Artworks and their Futures', in T. Anderberg, T. Nilstun, and I. Persson (eds.), *Aesthetic Distinction: Essays Presented to Göran Hermerèn* (Lund, 1988), 56–84, for more on this issue, and then compare what remains of his objectivist, if not his essentialist, view with anti-essentialist, hermeneutic, or deconstruction theories of meaning. One might consider Gadamer's suggestion that the meaning of an artwork is to be found not 'in' the artwork, but in its interpretative experience(s). Interpretation involves a 'fusion of horizons'—a fusion of a present with a past world-view to which the artwork gives us access. This fusion is mediated through the continuity of a tradition. On this account, an artwork takes on meaning through radically historicized and continuously revisable interpretation (*Truth and Method*). Gadamer has been criticized for making too strong a commitment to romanticism and to a conservative understanding of tradition. But there exist plenty of accounts of meaning in art not so committed. See R. Barthes, 'From Work to Text', *The Rustle of Language*, tr. R. Howard (Berkeley/Los Angeles, 1986), 56–64, and H. R. Jauss, *Toward an Aesthetic of Reception*, tr. T. Bahti (Minneapolis, 1982), ch. 1, esp. his argument against the 'platonizing dogma of philological metaphysics' (28).

being influenced by Brahms. One might also distinguish properties essential from an aesthetic point of view from those ontologically essential. Then one might identify a link between the two. In virtue of a work's being identified ontologically by its musico-historical context, it comes to have essential aesthetic properties provided by that context.[12] Levinson does not make use of such proposals, though he does synthesize ontological and aesthetic claims. Unfortunately, he leaves the synthesis unexplained; he seems content to treat ontological matters as theoretical—they must be argued for—and aesthetic ones in the guise of non-negotiable desiderata.

The relation between ontological conditions and aesthetic properties is notoriously difficult to specify, not least because the relata are in themselves problematic. But recall Goodman's solution that issues of quality have no bearing on issues of identity. He was motivated to hold this position, first, because when dealing with questions of identity ontological matters override all others; second, because he thought he had found a way to account for the identity of works without appeal to aesthetic notions. Even if Goodman implicitly took for granted some pre-ontological, or even some aesthetic, understanding of musical works, he was willing to dispense with such as soon as it conflicted with his ontology or could be replaced by an ontological understanding.

Levinson is more respectful of our pre-ontological and aesthetic understanding of musical works. He maintains that this sort of understanding underlies our very conception of musical works as peculiar phenomena that beg for ontological account. One must, therefore, try to reconcile theory with the understanding motivating the need for ontological theory in the first place. This sort of position is strengthened, however, only when the connection between ontological and pre-ontological and aesthetic understanding is made firm. As long as it remains tenuous, the more unpalatable, contrary position retains its credentials.

A fundamental conflict is once again demonstrated to exist in the analytic debate. Some theorists dispense with pre-ontological and/or

---

[12] Walton argues that certain 'facts about the origins of works of art have an essential role in criticism, . . . judgements rest upon them in an absolutely fundamental way' ('Categories of Art', *Philosophical Review*, 66 (1970), 337). He continues: 'The relevant historical facts are not merely useful aids to aesthetic judgement, they do not simply provide hints concerning what might be found in the work. Rather they help to *determine* what aesthetic properties the work *has*' (364). His use of the term 'has' appears to have essentialist import and if so, is problematic.

aesthetic knowledge where convenient, others try to accommodate it. This basic difference in aim seems to arise from a severe lack of agreement as to what counts as relevant to something's identity. For the remainder of this chapter we shall look, as we did in the previous one, at how serious the conflict is. We shall look at four cases, each of which will focus on one of the major claims of Levinson's theory and a contrary claim. Each time, we shall see that no satisfactory decision on the relevant matter can be made within the theoretical limits provided by current strategies of analysis.

VI

Case 1. If two composers independently produce original scores and performances, yet the scores turn out to be notationally identical and the performances 'sound the same', how many works have been created? In Levinson's view we have two works. Two (independent) composers means two works. In Goodman's view one work is created. That answer is inferred directly from his claim that notation is sufficient to determine the identity of the work in question. How does one decide the issue?[13]

Take a pre-ontological standpoint. Given two composers who independently produce identical scores, would we say unequivocally that they had produced different works? Surely yes. After all, knowledge of who composed a work seems to be an indispensable part of our knowledge of that work. The connection between composer and work is central to our very conception of musical works. Or surely no. Much of our evaluative discourse and musical analysis of works makes no reference to compositional source. If we hear performances that sound the same we immediately assume them to be of the same work. Notation plays a central role in this. It allows a work to transcend its specific compositional origin; it allows a work to acquire a certain amount of self-sufficiency so that it can come to be regarded independently of its maker.

Our involvement in musical practice seems to move in two directions—directions that reflect different aesthetic views as to what is

---

[13] See Wolterstorff, 'Towards an Ontology of Artworks', 137: 'A consequence of what we have been saying is that two people can compose the same work . . . Beethoven's *Opus 111* is not necessarily just an opus of Beethoven at all.' Cf. Walton, 'The Presentation and Portrayal of Sound Patterns', 238–9.

important and valuable about musical works. Upon what grounds could one now make a decision, first, to the effect that one aesthetic view is more adequate than another, second, that the preferred aesthetic view could be used as the basis for an account of the identity of the relevant phenomena? Without an answer to this two-pronged question the conflict cannot be resolved on either a pre-ontological or an ontological level.

Analysts think differently. It is precisely because there can be a conflict in our pre-ontological intuitions or between our aesthetic views that they look for independent theoretical arguments to justify ontological claims. But can these be found? If they can, will this not result in an unbridgeable chasm between the theory and what that theory purportedly seeks to explain? How precisely does the chasm appear?

Here is one way. On the level of phenomena and practice we have two existing aesthetic views. The resulting tension that exists, given contrasting views, is accommodated in practice because it exists in practice. It is accommodated if for no other reason than to provide variation and diversity in how we experience musical works. Are these different kinds of experience and the related criticism and evaluations to be encouraged? If they are, and if the tension is exciting and accommodated in practice, should the philosopher then strive to resolve or do away with the tension? A positive response to this last question generates the chasm alluded to.

Analysts might point out, and rightly so, that saying a practice accommodates a tension between different aesthetic theories is an insufficient explanation of that practice if, even in the context of practice, there is more to be said. And there is more, much more. Within musical practice there is a way of identifying and individuating works. We have decision procedures for determining how many works exist, procedures reflected in copyright laws and the like. The British copyright law of 1911 states, for example, that if two composers independently make use of the same tunes and produce a similar result, neither can be held liable to the other. If the second work sounds to the ear 'substantially similar' to the former, there will be 'a presumption of infringement'.[14] If there are decisions procedures in practice, all philosophers have to do is produce a theory that is compatible with the outcome of those procedures. Or perhaps such

---

[14] F. E. Skone James, 'Copyright', in Grove, *Dictionary*, 5th edn., ii. 433.

compatibility is not what philosophers strive after. The chasm reappears precisely when that conclusion is drawn.

In sum, whether we have one or two works is an issue resolved neither by simple consideration of alternative aesthetic theories nor by pure ontological arbitration. If a decision is to be made that coheres in theory and practice, we are going to need much more knowledge of the intricacies of the practice itself. Without it we cannot fully arbitrate between claims, unless arbitration means judging claims solely on the basis of their internal coherence and not on the basis of their accommodation of the existing practice.

VII

Case 2. Consider the claim made by Wolterstorff that musical works are not created, because the structures of works have always existed and thus pre-date compositional activity. Musical works, he continues, are products of compositional activity. But this activity is only a matter of discovering combinations of sounds and instrumental specifications and then of determining that these discoveries constitute the correctness conditions for performances. Only in discovering or selecting these combinations can the composer be said to be composing a work, because it is in this activity that certain combinations come to be constitutive of a work-composed-by-a-composer. Now Wolterstorff reminds us, there exist 'musical works that were probably never composed'. 'A composer does not bring that which is his work into existence . . . musical works exist everlasting'. Composition is not the same as creation. 'To compose is not to bring into existence what one composes. It is to bring it about that something becomes a work'.[15]

Whereas Levinson claims that the compositional act of conjoining a sound and performing-means structure is an act of true creation, Wolterstorff chooses to call this 'token creation' (the token creation of tokens)—an activity consisting of discovery and selection. Wolterstorff chooses this characterization because he sees (true) creation to be the bringing into existence of something that prior to the activity did not exist in any form at all—not even as inchoate matter. True creation is creation *ex nihilo*. True musical creation, to use Levinson's terminology, consists in bringing about both the initiated structures and the implicit structures from which the former are derived.

---

[15] 'Towards an Ontology of Artworks', 137; *Works and Worlds of Art*, 88–9.

Upon what grounds does one decide that composing does not count as creation? If we stipulate that creation is only to be understood as creation *ex nihilo*, we will be led to endorse Wolterstorff's brand of Platonism, or at least we shall be forced to conclude that to compose a musical work is not strictly speaking to create anything. However, if we stipulate that there are other ways of understanding creation—in terms of different forms of initiation—making, crafting, and constructing—we can accommodate the intuition that composers create their works.[16]

Can it be that the difference between the two theories ultimately turns upon a stipulation concerning the relation between composition and an ontological concept of creation, a stipulation that is, in neither case, fully explained? A stipulation like this is serious. For depending upon how one regards, say, the relation between creation and composition one is led to see works as having a certain kind of ontological status. Might it be that many disagreements among analysts rest on differently motivated stipulations, sometimes made out of allegiance to pre-established ontological positions, at other times out of allegiance to pre-critical evidence provided by the phenomena themselves?

VIII

Case 3. If a musical work is written for an instrument that ceases to exist before the work is performed and instrumentation is essential to a work, the work cannot be performed correctly. But surely, the

---

[16] In his 'Platonism in Music', Kivy argues for the discoverability condition on the grounds that identifying the work with a class of performances, the score, or an idea in the composer's head yields an unsatisfactory account. Consequently he feels, as he says, compelled to take a strong Platonic plunge. However, he also takes this plunge in the light of a useful discussion of creation and discovery, which, as he correctly shows, are more complicated and subtle notions than analysts' discussions typically suggest. Cox by contrast takes a Levinsonian sort of Platonic plunge. She argues in favour of the creativity condition: '(a) the number of elements and possible combinations of or relationships among elements available to a composer is virtually inexhaustible. It seems we are likely to refer to a work as a creation when the compositional process involves an infinite array of possibilities. (b) Most musical systems are artificial and dynamic. As a result, sonic elements have new and different implications in different historical contexts. (c) The style of a musical work is the result of human action, and is usually quite personal. The import of a discovery is that which is discovered, but the essence of a musical composition is a personal expression. These observations along with the fact that some musical compositions reflect or refer to perceptual phenomena of the material world, lead to the conclusion that musical compositions do not belong to the world of eternal being but to the world of temporal existence' ('Are Musical Works Discovered?', 373).

objection goes, there are works performed long after the specified instruments have ceased to exist.[17]

Specifications for instrumentation are usually broad enough to accommodate development and change of instrument types. When instruments go out of existence (and this happens rarely) something similar takes their place. Consider the development of the viol/violin and clavichord/pianoforte families. But can we reasonably eliminate a threatening objection by saying that something happens rarely or that the practice itself accommodates a certain tension? Not immediately.

One solution is to deny that instrumental specifications are essential to works, so that performances need not comply with them. This would not satisfy Levinson. He believes these specifications are not arbitrarily given to performers. Rather, they play a major role in determining the character of the work. Levinson has some history on his side.

Instruments (notably the human voice), from the earliest times, were believed to be crucial to sustaining a religious and human conception of music. Of course this information yields only a condition for something to be music: musical sounds must be produced by instruments. It does not motivate the condition that specific pieces of music have to be played on specific types of instruments. Precise instrumental specifications for individual works did become central to musical practice, however, as I pointed out earlier, in the late eighteenth century, concurrently with the so-called emancipation of instrumental music. Many composers at this time began to speak of instruments as 'individual personalities with voices'; many began to regard their specifications as unchangeable.[18]

[17] A distinction obtains between a work that *has not* as a matter of fact been performed and a work that *could not* be performed. Wolterstorff's comment that 'there are many unperformed works' misses this distinction (*Works and Worlds of Art*, 100). The challenge rests on the stronger claim that there are works that could not be performed. Incidentally, Goodman claims that instrumental specifications are essential to a work for rather specialized reasons having to do with notational redundancy (*Languages of Art*, 182).

[18] In a recent paper, Levinson strengthened his claims for the essentiality of instrumentation by arguing for a connection between each instrument (type) and an aesthetic quality. The aesthetic character of a work is determined in part by the instrumental specifications indicated at the time and in the original context of the work's creation. ('Authentic Performance and Performance Means') given at the Annual Meeting of the American Society of Aesthetics, 1988, Vancouver; pub. in his *Music, Art and Metaphysics* (Ithaca, 1990), 393–408. Probably Levinson would disagree with the conclusion drawn by the critic John Rockwell, who, in a recent review of some recordings of Bach's *Goldberg Variations*, considered the 'deeply complex issue . . .

Consider as well the practice of performing works on so-called period instruments. Part of its rationale derives from the belief that we acquire a more faithful and correct understanding of a given work if we perform the piece in the way it was originally performed at its time of composition. Such a view endorses the essentiality of instrumentation condition exactly in the spirit of Levinson's theory.[19]

But on further consideration we find a reason for thinking that instrumental specifications are not essential to works. If they are essential, how do we account for the belief that even though many works are transcribed or arranged for different combinations of instruments, subsequent performances are usually taken to be of the same work?[20] Works may be altered in many different ways and yet arguably remain the same works. There are *transcriptions*—cases of music originally written, say, for the violin now performed on the clarinet. There are *orchestrations*—cases where composers orchestrate works of other composers. Schoenberg orchestrated Brahms's Piano Quartet in G minor, Opus 25. There are also cases of composers orchestrating their own works. Schoenberg specified that his *Verklärte Nacht* be performed either by sextet or orchestra. Copland specified that his *Appalachian Spring* Suite be performed either by a small ensemble or full orchestra. Finally, there are *arrangements*, say, Busoni's and Kreisler's arrangements of music originally composed by Bach. Fritz Kreisler 'composed' almost exclusively by arrangement, or, 'in the style of'.[21]

---

about what constitutes music's essence. Is it', he asked, 'the abstract, formal structure of a piece, uncoloured by formal trappings? The Platonic noumenon, in other words, rather than a transitory phenomenon? Or is music itself subtly changed by the instruments on which it is played, and by the composer's and the performers' response to those instruments and the circumstances of the performance?' Rockwell concluded that despite the difference instruments can make to the performance of a piece, some pieces, and Bach's *Variations* is one of them, can 'seemingly thrive under almost any sonic conditions' (*New York Times* (25 June 1989), §2, pp. 27 and 42).

[19] Cf. Walton's remark that specifications of instruments have to do with how one performs a work in a way that is true to the historical context of the work. 'Performances of [a] Mozart piece', he writes, 'are to be heard as performances of an eighteenth century work' ('The Presentation and Portrayal of Sound Patterns', 255–7).

[20] One has to assume that practice is not simply begging the question against theory. If an adequate theory shows that one is not dealing with the same work in the case, say, of transcriptions, perhaps one should accommodate this insight in practice. Maybe one would say that a transcription of the work is being performed, but not the work itself, or maybe that a transcription is *directly*, while a work is *indirectly*, being performed.

[21] Kreisler often composed 'in the style of' past composers, though he claimed thereafter that the compositions were written by the composer in whose style he had written.

All these examples of instrument change, via orchestration, arrangement, and transcription, point to ways of producing what one might call *versions* of the same work. But to talk of versions of a work might be to contradict the claim that the work's original instrumental properties are essential. That claim implies that a work is individuated, at least in part, according to such specifications. So any tampering with instrumentation would amount not merely to a contingent alteration, but to a change (if such is possible) in the very essence of the work.[22]

Perhaps to produce a version is to produce a new work. Perhaps a transcription, an arrangement, or an orchestration of a work is itself a work in its own right, and to speak of it as a version has no ontological import.[23] Is this to go too far? Maybe only *some* kinds of instrumental change yield new works. Perhaps we should say that transcriptions do not yield new works, even though orchestrations and arrangements do. Liszt—the great transcriber—saw his transcriptions to be versions of the original work. He did not regard transcribing as a truly creative activity. In his recent biography of Liszt, Alan Walker records a claim once made by Liszt that he 'was the first to use the titles "Paraphrase," "Transcription," and "Reminiscence".' Walker follows with a note:

These evocative terms have become widely accepted, and Liszt ought to be given proper credit for introducing them. The paraphrase, as its name implies, is a free variation of the original. The transcription, on the other hand, is strict, literal, objective; it seeks to unfold the original work as accurately as possible.[24]

Perhaps orchestrations and arrangements do not yield new works either. When orchestrating Brahms's Quartet, did Schoenberg see himself to be creating an entirely new work? On the other hand, given

[22] Ziff uses the idea of versions differently: two 'works' that are audibly identical in performance yet are intentionally performances of works produced by different composers are called versions of a work, or one a version of the other. 'The Cow on the Roof', as part of 'The Aesthetics of Music', *Journal of Philosophy*, 70 (1973), 713–23.

[23] See S. Davies, 'Transcription, Authenticity and Performance', *British Journal of Aesthetics*, 28 (1988), 216–27: 'A transcription must depart far enough from the original to count as a distinct piece and not merely as a *copy* of the original' (217). None the less, the transcription should also 'resemble and preserve the musical content of the original work' (216).

[24] *Franz Liszt: The Virtuoso Years, 1811–1847* (Ithaca, 1987), 167 n. 12. Concerning Liszt's transcription of Berlioz's *Symphonie Fantastique*, Walker reports that Liszt 'bore the expense of printing his keyboard transcription himself, and he played it in public many times mainly to popularize the original score' (180).

the idiosyncratic nature of some of the orchestrations and arrangements that have been produced and, indeed, of many transcriptions as well, one might conclude for some cases of instrumental change (whatever type it is), but not for all cases, that new works are in fact created. Such a conclusion would surely please some musicians, but their pleasure is irrelevant here. The apposite point is that the question as to what counts as the production of a new work has revealed itself to be the sort of question that is unlikely to find a satisfactory answer merely through an ontological arbitration.

Levinson seems content to judge that transcription (he mentions no other kinds of instrumental change) involves the bringing about of new works.[25] More important maybe than its going against much historical evidence, this judgement counters his own gloriously romantic conception of composers. Schoenberg does not deserve his 'special glow' in virtue of his orchestration of Brahms's Quartet, however satisfactory the result. Kreisler and Glenn Gould are certainly not renowned for their compositional activities as such. Must Levinson now dispense with his intuitions about composers in the light of his ontological claim or can he hold onto both?

Levinson could have a way out of the problem. Transcriptions, orchestrations, and arrangements might be versions of musical works. If versions are not works, but only versions thereof, one need not mention them, let alone account for them, in a theory that makes no reference to forms of musical production other than work-production. The problem remains only if versions are works in their own right. Unfortunately Levinson does mention transcriptions and concludes that they are as much works as the originals upon which they are based.

There is another way out. Transcriptions etc. might not merit the same kinds of attributions as original works do, and as works, they might not fall under the concept of a work in quite the same way as the originals. Plausibly not all works share a perfect community of identifying properties. Is the work-concept more like a family resemblance concept? Does it admit a difference between paradigmatic and

---

[25] This has the unfortunate implication that Schoenberg's transcription of Brahms's Quartet is as distinct an entity from Brahms's work as it is from Beethoven's Fifth Symphony. Levinson might defend himself in Goodmanian style by suggesting that this is only an ontological implication. So, even if in aesthetic terms we appreciate the connection between the transcription and the original work, this has no bearing on the question of their respective identities.

borderline examples? If the latter, perhaps transcriptions are borderline examples of works.

How do these suggestions affect Levinson's theory? The answer depends upon whether his theory claims to account for all works or just a selection of musical works—perhaps paradigmatic examples. Levinson informs us accordingly:

I am confining my inquiry to that paradigm of a musical work, the fully notated 'classical' composition of Western culture, for example, Beethoven's Quintet for piano and winds in E-flat, Opus 16. So when I speak of 'musical work' in this paper it should be understood that I am speaking only of these paradigm musical works, and thus that all claims herein regarding musical works are to be construed with this implicit restriction. (6)

Presumably he does not intend to account for all examples of works. Why then does he even bring up the issue of transcriptions—or does he count them among paradigm examples?

Independently of Levinson's view, there are at least two conceptions of the role of instrumentation in classical music production, and both appear to be accommodated in practice. When composing, performing, and evaluating non-transcribed, non-arranged works, we typically treat instrumental specifications as essential. We respect a composer's specifications regarding instrumentation. When we engage in activities of transcribing and arranging, we suspend our belief in the importance of the original, instrumental specifications. However much we might see ourselves forced to decide the issue for ontological reasons, do we in any other terms want categorically to state that instrumentation is always essential to all musical works. In an important sense the answer depends upon what we think a musical work is.

IX

Case 4. This case concerns performances and the ontological accounts given of them. Many theorists, including Levinson, have described the identity conditions for a performance in non-extensional terms, and they have been motivated to do so given their recognition that Goodman's compliance condition has only a negligible chance of capturing how we use the concept of a performance in practice. Perfect compliance is a condition rarely if ever met outside the laboratory. Further, theorists have been unwilling to suspend the intuition that many performances containing mistakes are still to be regarded as genuine performances. They have also been unwilling to see performances merely as events compliant with notations. Unlike Goodman they do not accept that facts about the historical or human source of a performance are irrelevant to a performance's identity.

Imagine a situation in which three conductors, say Klemperer, Boult, and Beecham, each and in the right order conduct a different movement of a grand symphony. Would the resulting sound event constitute a performance of the work? In Goodman's view it would, granted only that the event perfectly complied with the score. How, where, and in what context a compliant is produced has no bearing on the identity of that compliant.[26] In a non-extensional view, by contrast, it is possible to countenance the idea that performances are individuated in part according to the work they express or interpret, in part according to their history of production. Performances, therefore, are not individuated with respect only to their internal (exhibited) properties—with respect only to what is heard.[27]

But what do expressive or interpretative factors have to do with identity? According to Goodman, how we account for a performance's identity is quite different from how we think about it qualitatively. No argument against Goodman is required here, however. Despite the

---

[26] An implication of Goodman's theory is that a performance can come about accidentally, say, by animals making certain kinds of sounds. A performance could be produced before the composer specifies the work's properties, giving us opportunity to remark that: 'the sound event we heard two years ago perfectly complies with this newly written score.' For further discussion of these kinds of consequences, see Walton, 'The Presentation and Portrayal of Sound Patterns'. To defend himself against these kinds of criticisms, Goodman appeals to the element of what is likely rather than what is possible and concludes that arguments based on unlikelihoods are 'utterly untenable' ('Comments on Wollheim's Paper', in Wollheim, Wiggins, and Goodman, 'Are the Criteria of Identity that Hold for a Work of Art in the Different Arts Aesthetically Relevant?', 50).

[27] Levinson distinguishes performances from instances or occurrences (27 n. 34). To be an instance, it is sufficient that the event perfectly comply with the score. But to be a performance, this condition is not necessary. Something is required other than perfect compliance, namely, that the sound event must be presented and recognized as a performance (expression or interpretation) of a work. Walton also argues that 'a performance not only presents a pattern, but portrays it in a certain light' ('The Presentation and Portrayal of Sound Patterns', 246). Sharpe argues, comparably, that what confers identity on a performance is not just that the same work is performed throughout, but also that a single interpretation is given throughout. In considering an example similar to mine, he suggests that when three conductors conduct different movements, one should say that we are hearing different parts of different performances ('Type, Token, Interpretation and Performance', 438). Kivy argues, contrarily, that it is counter-intuitive to discount the product of differently conducted movements from being a performance. He rests his argument on the claim that interpretative qualities deriving from how each conductor presents a work have no bearing on performance identity ('Platonism in Music', 122).

general motivation for providing a non-extensional account, the more formal accounts of performance identity need make no reference to qualitative features at all.

Recognize first a distinction between *internal* and *external* compliance of performance with score-copy. The former requires that a score's exhibited characters are complied with in a performance. By appealing only to the notation, internal compliance can be understood purely extensionally. *External* compliance, by contrast, involves non-extensional conditions. If we intend to produce a performance, say, in part by reading the relevant score, this is to intend-to-comply-internally. And that constitutes complying externally.[28]

In employing a notion of external compliance we effectively close the gap between a performance, *what* is produced, and the *production* of the performance. We commit ourselves (in a non-circular way) to seeing an intentional condition as contributing to the identity of a performance. Thus something is a performance of a work only if it is intended to be a performance of that work.[29] The intentional condition has to be understood as involving a certain amount of seriousness. Only someone with the requisite amount of skill, training, and knowledge could with any seriousness intend to perform a musical work. But the intentional condition, even if necessary, is insufficient to bring about a performance. Allowing for various contingencies one may with all seriousness intend to perform a work, yet fail to do this. To the intentional condition it is usual now to add a recognitional condition—a condition that speaks to the *reception* of a performance. Thus, something is a performance of a work if and only if (*a*) it was intended to be such a performance and (*b*) it is recognizable as an event intended to be a performance of this work.

Given these two conditions we can allow something to be a performance of a work even though it might admit of some wrong notes. But still these two conditions might not be jointly sufficient. If I intend to perform a work and find that what I play is recognized as a performance of that work, it is still possible for every note to have been incorrect. Internal compliance, in this case, would be entirely lacking. This might be the case if the performer and listener both thought that they

---

[28] The present formulation of the distinction is my own, but it functions in many non-extensional accounts.

[29] See Walton, 'The Presentation and Portrayal of Sound Patterns', 249: 'I propose that we consider a sound event a performance of a given work just in case its role, in the context in which it occurs, is to present the sound pattern identified with that work.'

were performing and recognizing a work by Beethoven. Yet, because of some long-term confusion, say, over the manuscripts, to which they alone had been subject, what they were actually performing was what everyone else thought was (and which internally complied with) a score by Haydn.[30]

This suggests that the individuation of performances should involve *some* internal compliance with the relevant score, so that, in addition to the intentional and recognitional conditions, we require 'enough', 'substantial', or 'significant' internal compliance with the score. Wolterstorff considers the idea of a performance 'coming fairly close' to instantiating all the normative properties of the work.[31] Thus something is a performance if and only if it is intended to be such, it is recognizable as such, and it comes fairly close to complying perfectly with the relevant score.

But how much is enough? Possibly we can sharpen our external compliance conditions to secure sufficient internal compliance. One must seriously intend to perform a work and this intention must stand in the right kind of relation to past performances or score-copies of the work—even in the right kind of relation to the original compositional activity. 'I intend to produce something which is like the performance I heard yesterday.' Alternatively, the relation could be a causal one. Walton writes, 'the causal history of a performance affects what it is a performance of.' The idea here is that a work's identity in a chain of performances, and thus also the identity of the performances, is guaranteed by the continuity of relatedness—be it intentional, causal, or both. Continuity serves to put and keep the performer and the listener on the right musical track.[32]

---

[30] This scenario reminds me of an announcement once made by Sir Thomas Beecham to his audience: 'Ladies and Gentlemen, in upwards of fifty years of concert-giving before the public, it has seldom been my good fortune to find the programme correctly printed. Tonight is no exception to the rule, and therefore, with your kind permission, we will now play you the piece which you think you have just heard' (H. Atkins and A. Newman (eds.), *Beecham Stories: Anecdotes, Sayings and Impressions of Sir Thomas Beecham* (New York, 1978), 29).

[31] *Works and Worlds of Art*, 86. Cf. Salmon's discussion of John Searle's cluster notion of the *Sinn* of a term, the denotation(s) of which has or have 'sufficiently many' of the relevant properties (*Reference and Essence*, 15 n. 9).

[32] Walton, Comments on Ziff's 'The Cow on the Roof', 726. Levinson writes that 'structural requirements aside, authenticity in . . . music is primarily a matter of intentional relatedness, not causal relatedness. The identity of a performance (say) is more a matter of what work the performer has in mind, of how he conceives of what he is doing—than of what work is the causal source of the score or memory which directly

Is this account adequate? Would we really want to say that if a person unintentionally plays something which internally complies with a score, and which we recognize to be a performance of the relevant work, we would not count this as a performance of that work because of the failure to meet the intentional condition? Possibly the intentional condition is neither necessary nor sufficient, just usual. Perhaps we would say that such a situation could never arise, or that we are playing around with the word 'intention', or that we have confused someone's having an intention with something's being intentional, or, finally, that we have confused intending to do something with being conscious of that intention. If we subtract the element of being conscious that we are doing something from the intention, do we not lose the point of the intentional condition? When finally we speak of intentions do we mean personalized intentions? The answer to this last question directly affects the multiple conductor case.

These sorts of questions can be asked interminably, and usually they have very little to do with musical performances. We are similarly sidetracked when we consider the condition of recognition. What does it mean for something to be recognizable as a given kind of thing, let alone as being a performance of a work? How do we recognize a work's first performance? But more important perhaps than any of these questions is that which asks whether, when the full picture of non-extensional conditions and causal chains is drawn, laden with provisos that there must be enough internal compliance, or that the intention must be successful to a 'reasonable degree', we find that the conditions have unduly constrained the nature of performance.

That question is more or less threatening depending on whether it is assumed that non-extensional conditions provide a sharp cut-off point between performances and non-performances. If it is acknowledged by contrast that these conditions are not to be interpreted too strictly, that they are designed to capture the essential vagueness or indeterminacy of what is still a successful procedure for performance-individuation, the criticism might lose some of its force. Theorists may have offered clues, but they have not offered definitive statements as to how their accounts should be interpreted.

So what are we to make of the multiple conductor case? Can we unequivocally say that the resulting sound event either is or is not a

guides the performer in producing the appropriate sound event.' ('Autographic and Allographic Art Revisited', *Philosophical Studies*, 38 (1980), 379. See also nn. 25 and 34 of same article.)

performance? Well it depends upon what one is willing to count as relevant to the identity of a performance. 'Of course,' analysts say, 'that is what we have been trying to determine all along.' Very well. But then all we need to know is what analysts mean when they speak of conditions relevant to identity. With this by now familiar comment we turn to the next chapter, where many of the critical threads already spun will be woven together.

3

The Limits of Analysis and the Need for History

It would be mistaken to assume that the use of the analytic method is identical in every theory of musical works, just as it would be mistaken were I to suggest that its particular use could easily be extracted from any theory. To further complicate matters, many principles and procedures of the method are taken on board without renewed consideration. Not every use of a principle can be justified in each and every piece of writing; there are always presuppositions. Still, there is something hidden in the method that forecloses the possibility of accounting satisfactorily for musical works. This chapter is devoted to describing what it is.

I have suggested many times that a basic tension in analysis regards the chasm it threatens to force between philosophical theory and practice. Is the chasm a necessary consequence of analysis, or just a contingent result of its perhaps inadequate use? If we get analysis right will the chasm disappear? This last question is complex because its answer depends upon what we regard as intrinsic and what dispensable to analysis. Were one to argue that the search, say, for identity conditions was misguided, could not analysts just acknowledge this and agree that the search be replaced with a more suitable one? To avoid unnecessary nit-picking over the 'proper' use of the term 'analysis', I shall adopt the following strategy. I shall isolate what I take to be problematic in some or all of the analytic theories as these have actually been presented in the literature. In each case, we will find that there is a methodological principle, a key assumption, or a central notion that is inadequately explained or unjustified. In each case, we will find that the inadequacy stems from an unsatisfactory or indecisive stand on the relation between the theory and the phenomena analysed.

The aspects of the problem I shall discuss are all closely related to one another. They are also not peculiar to theories of musical works, though I shall mostly articulate them as if they were.[1] This lack

---

[1] On occasion, I shall refer to art and artworks, and I shall do this to reflect instances

of peculiarity exists because problematic aspects tend to arise at a purely theoretical level before the application of the analytic method to any phenomena. Indeed, because they are so often at this level, the whole debate has a tendency to become entangled in metaphysical and scholastic complexities usually irrelevant to the phenomena at hand. Notwithstanding, I shall not consider the problematic aspects in all their metaphysical and scholastic complexity. That would constitute a book in itself. Rather, I shall presume many of their 'ins and outs' are familiar to readers. I shall only pick out and highlight in an exaggerated form those aspects apposite to our finding an adequate treatment of musical works.

I

Why do analysts seek the kinds of definitions for concepts that they do, and what, if any, explanatory power do these definitions have? Providing definitions has long been considered a respectable form of philosophizing. Definitions have been sought after for all concepts, be they natural or cultural, biological, physical, mathematical, moral, legal, or philosophical.

Reference to the history of musical concepts reveals at least one motivation for providing definitions in the musical sphere. The constant bid to define and redefine the concept of music derives from a need to convince the higher echelons of the establishment that certain musical practices are among those that are respectable and civilized. To establish the respectability of a given form of music one must make explicit what this kind of music involves as music. The debate carries on today with as much intensity as it always has. With the challenges of avant-garde musicians to what they conceive as outdated concepts of music, the question of classification remains an issue. With constant attacks on popular music, with its tendency, in Adorno's view, to constitute 'the dregs of music history', one is forced to seek knowledge of its musical nature.[2]

when the discussion of musical works has taken place only within the broader discussion of artworks.

[2] *Introduction to the Sociology of Music*, tr. E. B. Ashton (New York, 1976), 29. See also E. T. Cone, 'One Hundred Metronomes', *American Scholar*, 40 (1977), 443–59; J. Cage, *Silence: Lectures and Writings* (Middletown, Conn., 1961); and the opening paragraphs of P. Wicke's ' "Roll over Beethoven": New Experiences in Art', in his *Rock Music: Culture, Aesthetics and Sociology* (Cambridge, 1990), 1–2.

Analysts rarely seek definitions of musical or cultural concepts with these kinds of consideration in mind. Definitions are not usually sought after for reasons having anything directly to do with historical, political, or practical issues. What, then, motivates the search?

In the analytic tradition, under the manifold influence of the Enlightenment, Frege, and the Logical Positivists, the search for definitions has approximated as closely as possible to a scientific procedure, where the latter has been conceived in an anti-pragmatic manner. The dominant model of analysis for all areas of philosophy— ethics, aesthetics, and science—has come to be characterized as one governed by 'positivistic' standards of objectivity and logic.[3]

With the predominance of science, aesthetics has tended to stand on the borders of disrepute.[4] Aesthetic concepts, perhaps of all concepts, are believed to be least subject to scientifically styled or logical description. This belief has founded a tendency on the part of aestheticians to seek definitions, less because the definitions would be particularly useful for the understanding of the individual concepts themselves, but because it is a way of legitimating the discipline of aesthetics itself. Concepts and propositions associated with the arts are subject to the same kind of analysis as concepts found in the natural sciences, or so at least it seems to have been thought.[5]

II

A scientifically styled search for definitions has sometimes looked like an endorsement or modern extension of a traditional realist and essentialist metaphysics. Accordingly, it is claimed (*a*) that particular objects fall under a given kind (or concept) if and only if they possess the requisite essential properties; (*b*) if an object of kind K loses the

[3] Habermas referred to this approach as the scientization of humanistic study. See T. Bottomore, *The Frankfurt School* (London/New York, 1984), 18–19 and 27–34, for comparable and more detailed discussion.

[4] This was partly due to a particularly positivistic reading of the Tractarian distinction between saying and showing. It helped give rise to the idea that certain kinds of statement, especially ones about value, are not meaningful because they are not empirically verifiable.

[5] C. I. Lewis referred to aesthetics as a science. This is not an uncommon view nowadays, but Lewis still thought in his time that the science of aesthetics was quite undeveloped. See his *An Analysis of Knowledge and Valuation* (La Salle, Ill., 1946). See also Wolterstorff's discussion of natural kinds, norm kinds, and artworks as ontological allies (*Works and Worlds of Art*, pt. II, 46 ff.).

properties defining K's essence, it is no longer of kind K, and what is more, it is no longer *that* object at all because that object was K necessarily; (c) to provide a definition for K is to describe the essential properties associated with K; and (d) this definition holds for all time or at least as long as kind K exists.[6]

In line with this essentialism, concepts—even those functioning in cultural spheres—have been treated as fixed. A fixed concept is one that is unchangeable over time and can be described in terms of an immutable set of either essential properties or identity conditions. Evidence for this assumption comes less from explicit reference to fixed concepts than from employment of an ahistorical method. In the latter, one does not take the historical, contingent, and possibly changing character of the relevant concepts into theoretical account. Instead, one seeks to describe the purportedly essential content of any and all concepts—or, by extension, the pure ontological structure of any and all objects.

The extent to which essentialism has been repudiated within analysis is hard to determine. Different parts of the position are presupposed in the different theories of musical works; other parts are repudiated in various kinds of ways. Let us consider a modification that was once offered to a hardline essentialist view. It was suggested that if one wanted to suspend the search for essential properties, given their problematic nature, one could seek constitutive properties instead. These would be those properties that 'intuitively define the kind of thing it is, distinguish it from other closely related kinds of things, and in particular individuate the kind narrowly enough so that it can be intelligibly viewed as the possessor of the various entrenched aesthetic, artistic, and art-historical properties that musical criticism attributes to musical works'. Usually, it was added, the constitutive properties would turn out to be the essential properties associated with the kind in question.[7]

This proposal is problematic, and not only because it retains an essentialist commitment. If one wants to distinguish the search for

---

[6] Cf. B. A. Brody, *Identity and Essence* (Princeton, 1980), 135: ' "A has property P" means "a has P, a has always had P, there is no possible past in which a exists without P, and there is no moment of time at which a has had P and at which there is a possible future in which a exists without P." '

[7] This view was tentatively articulated by Levinson in comments on my paper 'Being True to the Work', given to the American Society for Aesthetics, Pacific Division, Apr. 1988. Goodman is excused from this part of the discussion, for reasons offered in his *Ways of Worldmaking* (Hassocks, Sussex, 1978), pts. 1 and 7.

constitutive properties from one for essential properties, the method and argument suitable for finding the former should be different from that suitable for finding the latter. One cannot continue to use the same arguments and just swap the word 'essential' for 'constitutive' to satisfy the critic of essentialism. To do so renders the distinction meaningless. Yet no alternative method was offered with the above proposal.

There are two responses one could make to this criticism, however. One might say that we have long been mistaken. We used to think we were finding essential properties, now we know better. We are really finding nominal essences in Lockean terms or constitutive properties as described above. That is why this terminological change does not necessitate, at least in the initial stages, a change in method or argument. Or, one might say that the traditional argumentation continues to be appropriate. It just has to be brought in line with a recognition that our concepts are in various ways related to our practices of performance and criticism. The former response still retains an essentialist commitment and its problems have been discussed extensively in the philosophical literature. The latter suggestion is more relevant to this essay, for it rests upon our seeking reconciliation between the search for essential properties and how we understand concepts to function in the relevant practices.

In my view, a reconciliation of this particular sort will forever evade our grasp. To see why that is so, we need to investigate the ramifications of attempted ahistoricity and, where applicable, related essentialist assumptions. These ramifications are diverse and far-reaching as many theorists—notably Ludwig Wittgenstein—have shown us. Let us explore just some of them, but not in familiar, general terms, but in terms pertaining to our comprehension of analytic theories of musical works.

III

Recognizing that many theorists treat concepts of the arts in an ahistorical and scientific way throws light on certain analytic claims and attitudes. The belief that such concepts are subject to scientifically styled analysis lies behind the claim, for example, that one virtue of producing an analytic theory is that it provides clear, unambiguous answers to questions in cases where no answers are given by the practice itself or by our pre-critical intuitions about that practice.[8]

Adorno spoke of such a rationale in terms of our seeking 'false clarity'. Perhaps there are good reasons why certain kinds of practice provide indeterminate answers to certain kinds of questions—questions having perhaps to do with whether or not instrumental specifications are essential properties of any or all musical works. Often questions like these, the analytic formulation of which demands a definite answer, ask too much of a practice that is indeterminate and complex. Should we then continue to provide answers of this sort?[9]

This negative response would be inappropriate if one believed that there was a natural order in musical practice which theory uncovers. It would be inappropriate if one believed, in other words, that revealing the underlying logic (grammar) of our language reveals the perhaps hidden ontological structure of that practice. For, given these beliefs, one might understand a theory of musical works to consist in a description of the logical form of our musical language (in which we speak of works, performances, and scores), which thereby describes the ontological structure of musical practice. If a theory is guided by these kinds of beliefs, the motto of analysis turns out to be a Russellian teaching, once also a Socratic teaching: do not be misled by how we commonsensically think and speak about things; do not be misled by pre-critical intuitions or our surface grammar. Unfortunately, it is not clear whether analysts treating musical works do actually hold these beliefs. To make matters more complicated, it is equally unclear whether they could hold any other underlying beliefs that would be consistent with the claims they make.

For Socrates, as for Aristotle, seeking definitions, or engaging in metaphysical speculation more generally, had this point. The search was inextricably bound to the idea of living a good life. It would help, through the knowledge or 'tethered' opinions yielded, to guide our actions. It would help us in the many situations we could find

[8] To undermine the tension implicit in his view that a transcription of a musical work yields an ontologically distinct work, Levinson claims it is a virtue of his view 'that it gives a clear answer to this question, which is often thought to be only arbitrarily decidable' ('What a Musical Work is', 27).

[9] Cf. Walton's line of questioning, beginning: 'These questions do not have clear answers. Is this because they do not much matter?' He then considers the suggestion that only if answers to questions matter to our critical and evaluative practices, do we have to find appropriate theoretical answers ('The Presentation and Portrayal of Sound Patterns', 240).

ourselves in to act appropriately and with good intention. It could help us also to understand the various practices and traditions constituting our society, thereby enabling us to change them (with greater or lesser success) if corrupt. The kind of clarity sought by analysts does not have this practical telos. It is rarely their aim to analyse the concept of a musical work in order to understand the practice of music within which that concept functions. It is also at this point that the Socratic–Russellian parallel collapses. The assumption that there is a direct correspondence between logic and ontology does not correspond to doing metaphysics for a practical end.[10]

One way to avoid practical problems and issues altogether is by sharply separating philosophical theory from all considerations of a practical sort. Goodman achieved this by employing strong provisos. The crucial proviso was also the one most heavily used, namely, that he refused to quibble 'about the proper use of words such as "notation," "score," and "work".' 'That matters', he continued, 'little more than the proper use of a fork.'[11] Rather, he claimed, all such terms are used theoretically in line with his philosophical concerns. After all, the idea is to produce a philosophical theory. No analyst other than Goodman seemed to be content with this proviso. Others wanted to 'quibble' over the ordinary usage of terms, at least in part.

But Goodman had a profound reason for adopting the position he did, that we are now at last in a position to make quite explicit. To justify his general position, Goodman noted first that definitions sometimes have to be stipulative. In an earlier book, *The Structure of Appearance*, he had argued that a stipulative definition 'is acceptable if it violates no manifest decision of ordinary usage. It can become legislative for instances where usage does not decide.' In *Languages of Art*, he wrote accordingly that

the problem of developing a notational system for an art like music amount[s] to the problem of arriving at a *real* definition of the notion of a musical work. Where a pertinent antecedent classification is lacking or is flouted, a notational language effects only an arbitrary, nominal definition of 'work' as if it were a

---

[10] This claim is somewhat exaggerated and is probably not true of early positivists. It has become decreasingly an exaggeration as theorists have increasingly lost sight of the original motivation for analysing concepts. I do not wish to undermine the need for clarity of argument and detailed theory, I only urge that a reasonable and useful perspective be retained whilst theorizing.

[11] *Languages of Art*, 189.

word newly coined. With no prototype, or no recognition of one, there are no material grounds for choosing one systematization rather than any other.[12]

When Goodman subsequently argued for the perfect compliance condition for musical performances, he stipulated that such a condition would legislate precisely what is and what is not a performance. Theoretical problems like the sorites argument persuaded him, furthermore, to produce a condition that manifestly violated or flouted ordinary usage of the term 'performance'. Presumably a 'pertinent antecedent classification', or at least an adequate prototype, was lacking in this case. That being so, Goodman's term 'performance' served in his theory as a 'newly coined word'. But few of Goodman's readers seemed to take it as serving just that role.

Goodman was all the time reminding us that on some level philosophical theory and practice can and should be considered separately, not because of some assumed correspondence between logic and ontology, but because what turns up in theory is not always (and should not always be) straightforwardly in conformity with practice.[13] But having said that, Goodman did not tell us how far he was willing to go. Was he asking readers to accept that once a nominal definition is offered, it should be considered forever more only within the context of the theory? Or was he perhaps suggesting that such a definition could have an effect on practice itself? I suspect the former, given the complications that arise when one suggests the latter.

The latter suggestion would make sense if one believed that philosophical theory had prescriptive force. If it did, one could claim that a well-worked-out or internally coherent theory could benefit a practice that lacked its own coherence. Following along, one could now consider whether in the light of a theoretical demand, say, for the condition of perfect compliance, we wanted to prescribe that condition for the practice in question. But that sort of consideration raises just the question that spurs Goodman to focus entirely on theory, namely, how does one judge the correctness or appropriateness of a theoretical condition?

In part, the answer depends on whether one wants to be faithful to a theory about a practice or to the practice itself, or to both. Suppose we decided to be faithful to a practice, and we took that to mean that we wanted practice to function either as it presently functions, or according

[12] *Structure of Appearance* (Cambridge, Mass., 1951), 5; *Languages of Art*, 197.
[13] *Languages of Art*, 220.

to a vision we had of what we wanted it to be like. In the particular case of musical practice we would probably reject Goodman's condition of perfect compliance as unduly restricting the production of performances—it is simply too strong a condition. Or we might not. Suppose, more generally, that we looked back at ordinary usage of the relevant terms to test the plausibility of the new condition. Probably again, we would reject Goodman's condition, or then again we might not. But these sorts of answers—intimated by Goodman's critics—would presuppose, *contra* Goodman, that ordinary usage or material grounds could be used to decide whether or not we accepted a condition formulated within a theory. More than that, we would be assuming that ontological theory, or at least parts of it, ultimately follows practical considerations, and, therefore, does not determine them. But what, then, of the theory's purported prescriptive force?

Suppose we arrived at the contrasting position which demanded that we follow theory whatever practical changes it incurred. And we were told to do this just because theory is more decisive, systematic, coherent, or whatever else. Now we would be openly acknowledging the prescriptive force of theory. But would we be any better off? I do not think so, for we would still be left with the problem of having to determine the specified condition's correctness. We would be left with the question, in other words, why we would choose this condition rather than that one. After all, there is more than one theory, and, within this particular scenario, we certainly could not look to material grounds, even if there were any, for deciding between alternative systematizations—or principles.

The point is this: as soon as we admit that a theory has any connection—descriptive or prescriptive—to practice, a basic dilemma arises: do we judge the adequacy of a theory, a definition, or a condition on theoretical grounds alone, or on the basis of the theory's subsequent accommodation in practice, or on the basis, somehow, of both? Not knowing where to go now, not knowing how to solve the dilemma, Goodman's solution beckons. Put in the strongest terms, it recommends that when we engage in philosophical theory we would do well not to touch the relation between theory and practice. Acknowledge the relation in principle, but then stay as far away from it as possible.

Not wanting to go to that extreme, one might decide to be less decisive, as Goodman's critics are. They do not ignore the problem, but they do not solve it either. They simply act on it indecisively, as if it were an inherent and unavoidable tension. One could of course embrace this tension enthusiastically. One might suggest that such a tension between theory and practice reveals the workings of a dialectic relation in which the relata—the theory and practice—constantly keep one another in check. But this suggestion is unlikely to appeal to theorists who have been tempted in the past to do away with all such 'Hegelian' tensions, in the name of clarity, determinateness, and neatness. This is unfortunate because the analytic theories stand to one another in positions of irreconcilable conflict, positions which place the conflict between theory and practice at the very core of the entire analytic debate. Put simply, no one can quite decide what sort of data or consideration is relevant to producing a philosophical theory of musical works.

IV

Treating cultural concepts scientifically or naturalistically works hand-in-hand with the assumption that the analytic method is free of certain complicated relationships, exactly of the sort existing, say, between the form a theory takes and the subject-matter it concerns itself with. For treating concepts in this way tallies with the belief that concepts are historically and ideologically neutral, unaffected by contingent changes, and undetermined in all 'essential' respects by those 'myths' or 'prejudices' inherent in our different cultural, social, political, and aesthetic milieux. That analysts treat concepts in this way derives from the general thought that culture is either something 'merely' transitory, arbitrary, and deeply ideological, thereby to be avoided at all costs, or something that approximates as far as it can to the state of nature or, in modern terms, to the ideal state of logic. Since ideology and ontology do not mix, one must—if one wants to produce an analysis, ontology, or metaphysics of culture—adopt the latter view.[14]

[14] There are more or less neutral ways to use the term 'ideology'. Terry Eagleton defines an ideology as a world view that helps determine particular social forces and structure power relations (*Literary Theory: An Introduction* (Minneapolis, 1983), 14). An ideology can also be defined more broadly as a world view comprising a way of life, a system of reference, a scale of value, and a set of mores and habits. To this list, borrowed from E. H. Gombrich's *In Search of Cultural History* (Oxford, 1969), 2, one might also add a conceptual scheme. I have benefited here from Gadamer's discussion of prejudice in his *Truth and Method*, 235 ff.; Peter Berger has also argued that with an emphasis on the reification of cultural phenomena, what is specifically human within what is essentially a human reality is ignored entirely (R. Wuthnow, R. J. D. Hunter, A. Bergesen, and E. Kurzweil *Cultural Analysis: The Work of Peter L. Berger, Mary Douglas, Michel Foucault, and Jürgen Habermas* (London/Boston, 1984), 75).

Whether once stripped of specific historical or ideological content a cultural concept has any content left is a moot point. To say it does might be to say that a concept has content (of whatever sort) belonging to it independently of the practice(s) in which it functions and that this can be known independently of knowledge of that practice. To say it does not would be to admit that some knowledge of the practice is required in an explication of the relevant concept. Again, even if there is basic acknowledgement of the latter point, it is inadequately accommodated within analytic theory.

Even the apparently harmless belief that all or many kinds of music can be spoken of in terms of works can serve as evidence for the inappropriate stance of analysis. Generally, it seems not to have occurred to theorists (at least until recently[15]) that the work-concept might not function in all musical practices of whatever sort. Thus where attention has been given to the concept of music (or the concept of art, more generally), it has automatically been given to the concept of a work of music (or art). Richard Wollheim opens his book *Art and its Objects* with the classic questions: ' "What is art?" "Art is the sum or totality of works of art." "What is a work of art?" ' But if we describe the particular kinds of work of art—painting, sculpture, music—are we automatically given an understanding of the more general concept of art? Wollheim says yes. But why?

Because Wollheim, like most others, has only taken logic into account here. If we want to consider a general concept, we can do this by examining the totality of particulars falling under that concept, and works of art fall under the concept of art. His reasoning rests on the assumption that the concept of art can automatically be understood with regard to the more specific concept of a work of art. The logical argument is fine. Yet, from a historical point of view, the transition from art to work of art is neither automatic nor just logical. The concept of a work of art, where this embraces, say, a work of music, a work of literature, or a painting, has not always been understood in the way it is today, and it stands in more than one relation to the different practices within which it functions. The relations between each art and its associated work-concept are not easily made subject to generalization, since in each case they are deeply historicized and are comprehended only by reference to the individual histories of the

[15] See F. E. Sparshott, 'Aesthetics of Music: Limits and Grounds' and Wolterstorff, 'The Work of Making a Work of Music', in P. Alperson (ed.), *What is Music? An Introduction to the Philosophy of Music* (New York, 1987), 33–98 and 102–29.

different arts. Questions, even ontological ones, about the concepts of art and music are considerably broader than those concerning the more specific concepts of a work of art and a work of music. None of these concepts should, therefore, be too quickly conflated or confused.

Perhaps, however, there is a way to translate questions about the concepts of art and music into questions about the concepts of works of art and music. If there is, an explanation is owed as to what is involved in the translation. Perhaps the translation is justified on the grounds that when analysing the concepts of art and music, the idea is just to analyse the concepts of 'fine' art and music. When theorists speak of the concept of a work of music, usually they confine their inquiries to the tradition of 'classical', 'serious', or 'concert' music, even if that does not stop them mentioning other sorts of music on the way. But that confinement has an historical explanation that we need to know about.

Just why is it in fact that theorists have chosen to focus on classical, concert music? For what reasons have ancient, Asian, folk, jazz, and popular forms of music mostly been left out of the inquiry? These are complicated questions, and they are made all the more so when we realize that even 'classical music' is a troublesome concept which, having entered musical language at the turn of the nineteenth century, has already taken on two distinct meanings: to denote music written in 'classical style' in the second half of the eighteenth century and to denote European 'concert' music in its entirety.[16] To focus just on classical music is to confine one's inquiry to a small, perhaps even to a very small, class of musical phenomena. And even if one could justify the focus on classical music, one would not thereby have justified one's focus on the concept of a musical work unless one had shown that a special connection exists between the idea of classical music and the work-concept. Analysts have not generally shown this, though sometimes they have assumed it. So the questions remain: why the focus on classical music? Why the focus on the concept of a musical work? And why not a focus on the concept of music in as broad a sense as it can be understood?

Perhaps, after all, the focus on the work-concept is justified independently of any historical consideration. Perhaps this concept in particular reveals just the kind of ontological complexity that is of real interest to analysts. Could this reason be extended outside ontology

[16] The idea of classical music entered into musical writings in the early 19th cent., as something to be contrasted with romantic music. I shall discuss this further in Ch. 9.

into aesthetics? Does analysing the work-concept rather than the larger concept of music more quickly lead one to interesting aesthetical problems? This latter question begins to make the discussion far too tenuous, the former less so. Anyway, it seems mostly if not entirely to be the ontological mutancy of works that has motivated the presentation of analytic theories.

V

And the same sort of critique applies to the way theorists have thought about the musical works as objects or, at least, ontologically. Amidst the deep concern with all the various ontological possibilities for speaking about musical works as objects, no one way of speaking has ever been supported out of prior and detailed consideration of what it means to think about musical works as objects in the first place. Yet there are questions that could be asked, answers to which would benefit an ontological inquiry.

For example, given the fact that music is a performance art, what are we to make of the fact that a significant part of it is manifestly about the creation, peformance, and reception of musical objects, namely works? What do we intend to capture in and about classical music practice when we speak about a work as something existing apart from its performances and scores? Do musicians involved in the production of all kinds of music produce musical objects? To what extent (and this is increasingly a popular question) should one stress the active over the productive side of the arts more generally, that is, the fact that they are peculiarly interesting human activities? Though there are reasons for concerning oneself with the fact that composers produce musical objects, other than performances and scores, analysts have not investigated what these reasons might be. In this sense, their ontological questions and answers stand alone and isolated.

It might be objected immediately that the questions I favour are historical or sociological at best, and have no significant bearing on ontological inquiry. If this is so, with what justification can answers to these questions be used in the guise of pre-critical subject-matter? One should not assume that if a given understanding is pre-critical on an ontological level, it is pre-critical on all levels. On any other level, it might show itself to be as critical as any understanding can be. And surely it is wrong to ignore partially or completely the critical nature of this understanding in order to preserve the internal, theoretical coherence or purity of a particular ontological method, or simply, as Goodman does, in order to generate a pure, philosophical theory.

VI

The traditional analytic use of paradigm examples is not without its problems either. For the use of such examples has depended not just on ambiguity of method but sometimes also on contradiction. When one produces definitions, one produces systematic descriptions of concepts or prescriptive stipulations as to their employment and signification. Either way there is a choice: one can formulate definitions on the basis of paradigm examples alone, or formulate them to reflect both the concept's paradigmatic extension and the boundaries of its entire extension. What is involved in making this choice?

One might confine one's inquiry to paradigm examples for this reason: by concentrating on undeniable specimens one more easily grasps the properties or 'distinctive traits' that the objects falling under the concept typically have. These properties can then perhaps be translated into essential properties. However, if we speak of typical or even constitutive properties, we are obliged to give a non-question-begging explanation as to why we pick out certain examples as exemplifying such properties. Empirical observation might help. But then properties that we posit as essential, constitutive, or typical in theory are apparently not established as such by empirical observation alone. Something like essential insight or metaphysical reasoning is required in addition.[17] That's the first problem.

The second and more serious problem is that when we speak in essentialist terms—or even in modified terms of constitutive or defining properties—do we not assume that all objects, however borderline, possess these properties, because it is in virtue of having them that objects are the kind of objects that they are? This last

---

[17] Cf. Ingarden's reasoning: 'In contemporary European music,' there is 'great variety, so that it is extremely difficult to formulate a theory of musical composition to satisfy all trends and also to reveal a structure or a cluster of properties common to all works. The contradictions between the various theories are for me less significant, because they are not on the whole concerned with the essential question of what the musical work really is, but with the question of what properties a musical work should possess, according to a particular artistic program, for it to be valuable as a work of art. . . . First we must solve the essential problem. . . . We shall take as examples such works *indisputably* belonging to music' (*The Work of Music*, 41–2). I have also benefited in this discussion from S. P. Schwartz (ed.), *Naming, Necessity, and Natural Kinds* (Ithaca, 1977).

question points to a hovering contradiction in any theory that relies on the positing of defining properties, yet is purposefully confined to paradigm examples.

Perhaps the confinement is strategic: one could focus provisionally on paradigm examples and leave others to extend the account to all examples or to new examples if the situation arose where this seemed necessary. Whether this is the line actually taken by any theorist is unclear. But the lack of clarity leaves us in the dark. Just why is it that no theory of musical works has ever contained any explanation (*a*) for the examples it uses, (*b*) for its confinement to paradigm examples (if that approach is taken), or (*c*) for what counts as a borderline example? The answer is rather illuminating.

Theories of musical works have been formulated on the basis of examples drawn from the classical repertoire of the early nineteenth century. Beethoven is the composer, and his Fifth Symphony the work, most frequently referred to. Some theorists deliberately confine themselves to examples from this repertoire. We are not told why. Perhaps they are engaging in what Lakatos once referred to as 'monster barring'—the exclusion of 'difficult' examples? Others seem to want to account for all examples, however difficult.[18]

But it is revealing that examples drawn from early music, avant-garde music, often from folk, jazz, and popular music, but rarely if ever from the music of the nineteenth-century classical repertoire, are appealed to especially when one theorist challenges a definition offered by another theorist. 'It appears that Ingarden's elitism has a serious consequence for his analysis of the musical work,' his translator Czerniawski writes. 'By ignoring popular works, he fails to realize that the ontic status of a musical work is variable, since at the popular end . . . its identity is uncomplicated by any score/performance relationship.'[19] The point could not be more succinctly shown.

[18] Cf. I. Lakatos's discussion of 'monster barring' in his *Mathematics, Science, and Epistemology* ed. J. Worrall and G. Currie (London/New York, 1978), 15–16.

[19] *The Work of Music*, p. xiv. Recall Tormey's examples of aleatoric music and Ziff's examples of early music used against Goodman's theory; also Levinson's remarks on transcriptions. Consider, too, Wolterstorff's comments on improvisations: on p. 74 in *Works and Worlds of Art*, he remarks that with an improvisation 'there is no performance of a *work*'. The exclusion is based on the lack of there being a predetermined or pre-composed work. But on p. 105, he uses improvisations to undermine Goodman's theory: Improvisations, he says, fail to meet the retrievability test, 'for we do not yet know what in the performance is to be optional with respect to correctness and what is to be required'. Here, he seems to be presuming that improvisations are works, since he uses them as counter-examples to a theory of works. At other places, he speaks of folk

The success of a challenge using examples from all these non-classical sorts of music —the success of the counter-example method—depends on whether the examples picked out are examples of works, be they paradigmatic or borderline. The challenge rests on the assumption that the work-concept can be employed when speaking about all these kinds of music. Even if a correct assumption, it is a big one. Yet few recognize this assumption let alone provide any adequate explanation for it.

Theorists have wanted to have it both ways: first, they have wanted to consider their theories as based on works such as Beethoven's Fifth Symphony. Then, they have wanted to use examples given from early music, avant-garde music, and other sorts of music, either to knock down their opponent's theories, or, more charitably, to acknowledge in their own accounts that forms of music other than post-1800, classical music do exist. This procedure has gone on without comment. Yet it is a serious issue how we chose examples as subjects for theoretical account. It is just as serious an issue how examples come to be considered paradigmatic of a concept. For when we inquire into these matters it turns out that some central procedures of the analytic method have to be rethought and perhaps replaced. One such procedure is the search for identity conditions, another is the traditional counter-example method.

VII

Consider, for now, the search for identity conditions. (I shall leave my remarks about the counter-example method until much later, as they follow naturally on from the historical remarks.) In seeking to describe the status and identity of musical works a particular style of reasoning

music in terms of works (67 and 76). In folk music 'the work just emerged from performances' ('Towards an Ontology of Artworks', 67). Even though, in oral traditions, there are no conventions for producing scores, apparently works are still produced. Ingarden also speaks both of folk music and improvisations in terms of works (*The Work of Music*, 38). But when Ingarden's critic, Zofia Lissa, used examples of recent avant-garde music against him, he responded wryly that he could not account for such examples in his theory because at the time of writing (1928) they were not yet in existence. 'Phänomenologen sind keine Propheten und wollen es auch nicht sein.' He had deliberately confined his remarks on musical works to a given range of paradigm examples. See Lissa, 'Some Remarks on the Ingardenian Theory of the Musical Work', in P. Graf and S. Krzemien-Ojak (eds. and trs.), *Roman Ingarden and Contemporary Polish Aesthetics* (Warsaw, 1975), 129–44, and Ingarden, 'Bemerkungen zu den Bemerkungen von Professor Zofia Lissa', *Studia Filozoficzne*, 4 (1970), 351–63.

has often been adopted, a step-by-step progression from the specification of what are taken to be the essential properties of works or their identity conditions to a conclusion about their ontological status. Suppose we think the fact that a work is composed indicates something essential about it. Then we shall see it as having an ontological status embracing the fact that it was created. The idea that the work, say, is a universal, conceived as something eternally existing, would thereby be excluded. The connection made between the specification of identity conditions and the determination of ontological status is therefore crucial, for success in the latter depends upon that in the former. It is, however, problematic, and for many reasons.

The most important problem arises not on a purely philosophical level, but at the intersection between philosophical and other sorts of considerations. As we have already shown, it is simply not clear how one decides what counts in any given case as an identity condition, and what type of justification can be offered for one. There seem, in more specific terms, to be no clear grounds for deciding whether purported conditions for a work's identity (or properties purportedly essential to a work) are what they are claimed to be, i.e. essential.

Consider the argument to the effect that the compositional condition has no bearing on the identity of a work, even if knowledge of the composer is relevant to our aesthetic appreciation of, or even in our practical dealings with, this work. Knowledge of the composer might help us to better perform or listen to the work more adequately, but this has little to do with its identity as such. *A fortiori*, though the instrumentation specified by composers is aesthetically relevant to their works, that the works are performed on the specified instruments has no bearing on their identity.

A distinction between ontological and aesthetic considerations is operating here, a distinction that admits the possibility that a certain property can be considered irrelevant from an aesthetic point of view, yet none the less be regarded as essential ontologically, or vice versa. To see these considerations as separable is not necessarily to see them as disconnected. But if a connection is assumed it has to be demonstrated clearly. Contrarily, if no significant connection is presupposed, one is led to ask what grounds are used to guarantee that a property is ontologically essential? As we asked before, could the internal demands of a given ontological theory ever suffice?[20]

---

[20] See Wollheim, 'Are the Criteria of Identity that hold for a Work of Art in the different Arts Aesthetically Relevant?', 29–68; and Strawson, who directly suggests

One danger of severing issues of identity from aesthetic considerations, or indeed from considerations of any non-philosophical or non-logical sort, is that an appeal to the latter will not even be able to help, let alone resolve, theoretical disagreements. Not everyone thinks this is a problem. Still the lurking danger remains that the theories will probably become forever divorced from the phenomena and practices they purportedly seek to explain, as well as from any non-philosophical interest we have in those phenomena. The problem with the search for identity conditions resides just at this point, then, in the incompatibility between the theoretical demands of identity conditions and the phenomena to be accounted for.

VIII

Analysts who have sought to describe musical works have employed methodological principles and assumptions that impose unnecessarily severe limitations on their theories. The fact that analysis has been designed not to treat different sorts of subject-matter, but rather to capture only the pure ontological character—the so-called 'logic'—of any given phenomenon, turns out to be the source of all its trouble. For this design has created an irresolvable conflict between theory and practice. While the analytic method has given theorists a way to account for the logic of phenomena, this has not been true for their empirical, historical, and, where relevant, their aesthetic character. Some theorists have simply ignored the complex character of the phenomena, while others have found that they could not, and furthermore that they have not wanted to. So the latter have sought reconciliation between the two. Reconciliations that have been found have been uncomfortable and ultimately unsatisfactory. Analysis has just not produced the methodological tools to make them otherwise.

that 'the criterion of identity of a work of art is the totality of features which are relevant to its aesthetic appraisal' ('Aesthetic Appraisal and Works of Art', 185).

PART II
THE HISTORICAL
APPROACH

4

The Central Claim

How does musical practice operate and how does the work-concept operate within it? The purpose of this chapter is to identify the philosophical content of the claim that the work-concept began to regulate a practice at a particular point in time. One way to do this is to investigate matters with an eye not just to ontological puzzles, but also to how a practice lives and survives—for indeed it does—without explicit understanding of its ontological structure. This change in emphasis involves stepping in all kinds of new directions, new at least to those who have been predominantly interested in analysis.

The main difference between the new approach and analysis is the explicit use it makes of history. This approach does not obviate the need for ontology, however; ontology is just reconceived to become inextricably tied to history. This does not mean either that ontological claims are justified solely by history or that the claims are incoherent on their own terms. It means a methodological priority is given to making ontological claims compatible with the historical and conceptual complexity of the subject-matter with which they are associated.

The proposed inquiry is designed to resolve or at least to change the terms of certain kinds of conflicts. Recall the conflict between Goodman and Ziff. In assessing the force of Ziff's purported counter-example to Goodman's theory, the question was posed whether ornaments affect either the quality or the identity of Tartini's 'Devil's Trill' Sonata, or both. It was also asked whether avant-garde musical production yields works in all or even some cases. The new strategy is to investigate how far the conception implicit in the production of early music, and then of avant-garde music, matches that implicit in work-production. Of course we have to establish what the production of musical works involves first, but we can and shall do that. But before doing any of this, I need to make explicit the ontological picture to be presupposed in the historical investigation.

The ontological picture comprises claims about concepts and objects that together constitute an account of the normative structure of classical music practice. I shall claim specifically of the concept of a musical work (i) that it is an *open* concept with *original* and *derivative* employment; (ii) that it is correlated to the *ideals* of a practice; (iii) that it is a *regulative* concept; (iv) that it is *projective*; and (v) that it is an *emergent* concept. I shall treat each in turn; after that I shall discuss the central historical claim.

I

The major methodological transition is a move away from asking what kind of *object* a musical work is, to asking what kind of *concept* the work-concept is. An answer to the latter will eventually yield a new answer to the former. But consider, first, a well-known theory of concepts, usually referred to as the theory of open concepts, despite its also describing closed concepts.

Despite anticipations in Nietzsche, the idea of an open concept—often also referred to as an open-textured concept—was first explicitly used by Friedrich Waismann in his essay on the verifiability of empirical statements. Waismann's account is reminiscent of Wittgenstein's remarks on unbounded concepts and family resemblances, and is well, if not exhaustively, understood in the light of these. Concepts of democracy, justice, and art, as well as of music and musical work, are examples to keep in mind as we proceed with the more general discussion.[1]

The theory rests upon three premisses. First, the different roles concepts have within a practice, the different uses to which they are

---

[1] Cf. Nietzsche: 'As for the other element in punishment, the fluid element, its "meaning" in a very late condition of culture . . . the concept "punishment" possesses in fact not *one* meaning but a whole synthesis of "meanings": the previous history of punishment in general, the history of its employment for the most various purposes, finally crystallizes into a kind of unity that is hard to disentangle, hard to analyze and, as must be emphasized especially, total *indefinable*. (Today it is impossible to say for certain *why* people are really punished: all concepts in which an entire process is semiotically concentrated elude definition; only that which has no history is definable.)' (*Basic Writings of Nietzsche*, tr. and ed. W. Kaufmann (New York, 1968), 'On the Genealogy of Morals', Second Essay, §13, 515–16.) Waismann, 'Verifiability', *Aristotelian Society Proceedings*, supp. vol. 19 (1945), 119–50; Wittgenstein, *Philosophical Investigations*, tr. G. E. M. Anscombe (Oxford, 1958), §§ 67 ff. Since the 1940s, the idea of an open concept has been employed widely in aesthetics, ethics, and the philosophies of science and law. See M. Weitz, 'The Role of Theory in Aesthetics', *Journal of Aesthetics and Art Criticism*, 15 (1956), 27–35; M. Mandelbaum, 'Family Resemblances and Generalizations concerning the Arts', *American Philosophical Quarterly*, 2 (1965), 219–28; W. B. Gallie, *Philosophy and the Historical Understanding* (London, 1964), ch. 8; T. J. Diffey, 'Essentialism and the Definition of "Art" ', *British Journal of Aesthetics*, 13 (1973), 103–20; H. L. A. Hart, 'The Ascription of Responsibility and Rights', *Proceedings of the Aristotelian Society*, 49 (1948–9), 179–94.

put, determine their character. Whether a concept is open, or by contrast closed, has less to do with the logic of concepts *per se* than with its multifarious possible uses.[2] Second, to account satisfactorily for the distinction between open and closed concepts depends upon one's rejecting a traditional, essentialist or realist theory of meaning. It depends instead on the construction of a theory that takes seriously the idea that human beings have decisionary power and control over their language and concepts. Third, the meaning of concepts cannot be analysed independently of the practice in which they function, since they acquire their meaning just by functioning in particular ways within practices. And since practices are not known or learnt about a priori, knowledge of conceptual meaning can be no different.

These premises are consistent with a particular view of language. Natural languages are not fully determinate, and the rules governing them are not fully circumscribed. Objects do not fall under a given concept simply or just because they share common or essential, exhibited or non-exhibited properties (or, in Goodman's terms, a perfect community of properties).[3] Rather, concepts and their extensions are adaptable according to their role in activities and the theories bound up therewith.

Open concepts have most often been described as:

(i) not corresponding to fixed or static essences;

(ii) not admitting of 'absolutely precise' definitions of the sort traditionally given in terms of necessary and sufficient conditions;

(iii) intensionally incomplete or 'essentially contestable'—because the possibility of an unforeseen situation arising which would lead us to modify our definition can never be eliminated;

(iv) distinct from, though related to, vague concepts. According to Waismann a concept is vague if there are cases in which there is no definite answer whether the term applies. ('Pink', 'tall',

---

[2] I am sympathetic to Hart's suggestion that the open texture of legal concepts is to be handled by extra-theoretic notions.

[3] Recall Mandelbaum's criticism of Wittgenstein's family resemblance doctrine. Mandelbaum argued that family resemblances function, if at all, with respect only to the exhibited (internal) properties of objects. We only fail to notice common features of different objects called by the same name if we attend just to their exhibited features. If we attend to external, relational features, such as genetic connections, common features can be found. The distinction between internal and external properties is notoriously difficult to formulate, but it often marks the difference between monadic and relational properties. Here, it suffices to note that it is possible to see a concept as closed with respect either to its internal or external properties. The distinction between internal and external properties cuts across that between open and closed concepts.

'bald', and 'middle-aged' are examples.) Open texture provides for both the logical and empirical 'possibility of vagueness'.[4]

But how, in the light of these descriptions, are we to understand closed concepts? Are they to be understood, *contra* (i) above, in essentialist or realist terms? No. That would commit us to a view of meaning whose rejection gave way to the doctrine of open concepts in the first place. A distinction must be drawn between fixed and closed concepts. Only the former function or find their expression within an essentialist view of language and the world. The doctrine of open *and* closed concepts, by contrast, is designed to sustain a non-essentialist view.[5]

Are closed concepts those expressed within a (Carnapian) formal, or what has been called an ideal or exact, language? No, to the extent that this understanding leads us to deny there are concepts, such as 'quart' and 'freshman', that are closed yet do not belong to what we usually call a formal language. Still, we are moving towards a plausible description. Thus closed concepts function within systems or practices requiring different kinds of formality, exactness, or precision. While mathematical and logical systems are obvious formal languages, measurement and monetary systems, the House of Commons, Cambridge University, and the Royal Family are precise or formal systems of a rather different sort.[6]

Concepts, perhaps going by the same name, can be treated as closed for certain purposes, and as open for others. For funding and insurance purposes, say, as opposed to purposes of criticism and aesthetic experience, certain concepts (often relating to ownership) most effectively function as closed. Whenever a concept is treated as closed, it is given exact and complete definition in the light of a stipulation made at a given time 'for a special purpose'.[7] This definition stipulates boundary conditions. In closing a concept we decide it is to be used if and only if the relevant objects have certain features. We recognize

---

[4] Recall my comments on vagueness in Ch. 1.

[5] Recall my claim that analysts tend to treat the work-concept as fixed. In this context, Dewey's insight is appropriate, that looking for concepts with hard and rigid boundaries is connected to the philosophical search for certainty, as argued in his *The Quest for Certainty: A Study of the Relation of Knowledge and Action* (New York, 1929), ch. 2.

[6] The best examples of closed concepts come from the sciences: the concept of anaemia is closed by the condition of a person's having a low blood count. The concept of triangle is closed within the Euclidean system.

[7] Wittgenstein, *Philosophical Investigations*, § 69.

the stipulation is dependent upon the use to which we want to put the concept. Thus when we want to change the system and thereby the use of the concept, we change the definition. Though we might continue to use the same name, we use a new concept because we give up the old definition and replace it with a new one.

Open concepts are different. Because they function within a different kind or part of a practice their definition does not require a stipulation of boundary conditions. When we treat a concept as open we treat it as unbounded; its definition need be confined only to known or uncontroversial, canonical, or paradigm examples.[8] Open concepts are treated so that they can undergo alteration in their definition without losing their identity as new examples come to appear as standard, as the practice within which they function changes. Unlike definitions of closed concepts, those of open concepts are expanded and modified but not replaced. Open definitions, if one may call them that, are not treated then as rigid or fixed, but as 'signposts' facilitating language use. They are mutable and flexible in the light of their particular descriptive and prescriptive functions.[9]

Continuity is crucial to the functioning of open concepts. The continuity weaving through a living and changing practice—say, an artistic or moral practice—often manifests itself in the continuity of function of its open concepts. Such continuity is guaranteed through the expansion or modification of definitions rather than through their replacement. This prompts us to trace the genealogy of the concept or the history of its meaning as it has functioned within the relevant practice as a way to understand both the concept and the associated practice. With closed concepts our consideration has a different

---

[8] Ibid., § 68: 'I *can* give the concept "number" rigid limits, . . . that is, use the word "number" for a rigidly limited concept, but I can also use it so that the extension of the concept is not closed by a frontier. And this is how we do use the word "game". For how is the concept of a game bounded? What still counts as a game and what no longer does? Can you give the boundary? No. You can *draw* one; for none has so far been drawn. (But that never troubled you before when you used the word "game".)'

[9] Recall Wittgenstein's idea that a rule cannot be made to cover all possible future applications of a concept, but this does not imply that serviceable rules cannot be given for the present (ibid., § 84). For mention of serviceable definitions, see Margolis: 'Definitions may serve only to fix the properties of what, in accordance with the prevailing current of competing theories, are thought to be normal or central instances. . . . Definitions are practical and alterable instruments servicing developing theories that cover at least certain undeniable specimens.' (Margolis (ed.), *Philosophy Looks at the Arts: Contemporary Readings in Aesthetics*, 3rd edn. (Philadelphia, 1987), 139.) For comparable, but more general, discussion, see H. Putnam, 'The Meaning of "Meaning" ', *Collected Papers*, ii (Cambridge, 1975), 215–71.

emphasis. We assess the practice or system less in historical terms than in terms of its own particular internal, formal, structural, or purposeful coherence—especially when we are confronting issues of justice, medicine, or property.

An open concept sometimes undergoes quite radical shifts in function and meaning, but it does not thereby lose its identity. Its identity is preserved by the continuity that is guaranteed at any point in time if the concept's present use is appropriately connected to its previous uses. These connections, in formal terms, are causal, intentional, and recognitional, but in practical terms, the connections are shown through examples, say, of a past event inspiring the production of a future event, or of a desire to develop, expand, or improve upon something done in the past—the sense of working within a tradition. The presence of continuity or connectedness also allows us to explain how a given object may at one time fall under an open concept, and at another time not do so. Knowledge of connectedness enables one to recognize that a given object once fell under a concept and to know why it no longer does, if indeed it no longer does.[10]

We should never be misled by the idea of openness. We might think that calling a concept open implies that 'anything goes', that unless a concept has a fixed or bounded identity, anything in principle could fall under it. This is mistaken. As Ronald Dworkin puts it: 'Discretion, like the hole in a doughnut, does not exist except as an area left open by a surrounding belt of restriction'.[11] Conceptual change is restricted in precise and different ways depending upon a concept's particular use.

Suppose you wanted to produce a radical change in the way a given concept was used, and you thought this could be done by producing something incorporating a denial of everything we (others) thought was involved in the employment of this concept. We could only understand the result as a use of *this* concept if we found it to have a significant connection to something which constituted a use of this concept in the past. Otherwise we would simply deny your employment had anything to do with the concept at all. Not anything goes. But this does not mean that nothing goes, or even that only a little goes.

[10] Cf. Wittgenstein, *Philosophical Investigations*, §67: 'Why do we call something a "number"? Well, perhaps because it has a—direct—relationship with several things that have hitherto been called number; and this can be said to give it an indirect relationship to other things we call the same name.'

[11] 'Is Law a System of Rules?', in Dworkin (ed.), *The Philosophy of Law* (Oxford, 1987), 52.

If this is a plausible view of concepts, and I think it is, the nature of conceptual continuity and relevant restrictions must be spelled out carefully from case to case. The claim that a concept is open can, therefore, be no more than the beginning of the story, certainly never the end.[12]

II

The theory of open concepts tells us something about both our practical *and* theoretical employment of paradigm examples. Consider Morris Weitz's claim that, even though there is no single fixed set of defining properties, say, for works of art, there are still paradigm examples of them. These are chosen on the basis of criteria, he argues, such as of recognition and evaluation, determined in turn by prevailing, dominant, and competing aesthetic theories. Weitz is renowned for having claimed that art is an open concept.[13]

Ziff took another step. He claimed that there is neither a single set of defining properties for something to be art, nor are there paradigm or 'clear-cut' examples of art. Looking back into the long history of art, and at the many uses the concept of art has had, at which point would we say that the art being produced was paradigmatic of the tradition? Ziff combined the two claims by arguing that one reason we have for not being able to find an essentialist definition for 'art' is precisely because there are no paradigm examples.[14]

Are paradigm examples to be understood, as Ziff intimates, in accordance with an essentialist view of things? Are we to assume that a given example is *essentially* a paradigm example, such that what we take to be paradigmatic of a tradition cannot change as the theories and corresponding criteria of recognition and evaluation change? I do not think so. In contrast, it is possible, though our employment of a concept is multifarious and changing, to find paradigm examples of that concept. They are paradigmatic, not because they are associated

[12] This discussion is compatible with other theories about the 'revisionary' character of practices. I am thinking here especially of Peirce's and Dewey's conceptions of pragmatic and instrumentalist inquiry and Gadamer's conception of hermeneutical inquiry. Cf., for example, Gadamer's remark: 'Changing the established forms is no less a kind of connection with the tradition than defending the established forms. Tradition exists only in constant alteration.' (Quoted in D. C. Hoy, *The Critical Circle: Literature, History, and Philosophical Hermeneutics* (Berkeley/Los Angeles, 1982), 127.)

[13] 'The Role of Theory in Aesthetics', 28.

[14] 'The Task of Defining a Work of Art', *Philosophical Review*, 63 (1953), 68–78.

with an unchanging set of essential properties, but because—to turn Weitz's point back into Wittgenstein's—they play, for a given time, a particular role in the practice in which they exist.

Certain examples falling under a given concept are paradigmatic because they are chosen by us according to some purpose we have. Given this purpose, an example is duly accorded paradigmatic status. This is only a general principle, and a vague one at that.[15] By itself it does not tell us why we actually choose the paradigm examples we do. Part of an explanation for that is, however, forthcoming.

Take the concept of a musical work. At a particular point in history, this concept attained a centrality, a certain kind of status in musical practice, and it acquired this status just because of the particular way in which it emerged. This *institutionalized centrality* is, in my view, the foundation for our acquiring at a given time a standard or model by which we choose certain examples as paradigmatic. That we might continue thereafter to use the same paradigm examples is then explained by the further use we make of one and the same standard. Institutionalized centrality is closely related to what in more familiar terms we identify as a mainstream or a canon. The emergence of a mainstream—housing the paradigmatic—depends upon a practice's being standardized with respect to various aspects of its structure.

The principle of institutionalized centrality contrasts with a principle according to which our choice is determined by what we like best, or by what is produced contemporaneously or nearest to hand. It also contrasts, though it is not incompatible with, psychologically based principles. There are many other such principles. I believe, however, that we have tended, and still tend, to pick as our examples those works produced at the time when the work-concept acquired a centralized position in musical practice. For reasons that shall become clear later on, we tend more often than not to choose works by Beethoven.

All these general suggestions fit concepts and practices of many different sorts. But in any given domain they point to the insufficiency of concluding that our choice of paradigm examples is dictated according to needs and purposes, unless we also show how these needs and purposes are connected to the historical character of the practice.

[15] Cf. Ted Cohen's remarks on the problem of authoritative legitimation in the art institution which forms part of his discussion of the 'institutional theory of art' ('The Possibility of Art: Remarks on a Proposal by Dickie', *Philosophical Review*, 82 (1973), 69–82).

In classical music practice, for example, one has to look at the actual uses to which the work-concept has been and can be put, to discover the needs and purposes existing within it. Then one discovers how these needs determine our choice of paradigm examples.

Explicating the work-concept historically does not require us, however, to confine our inquiry to paradigm examples. On the contrary; to understand why we choose certain examples as paradigmatic, we must also understand how examples can be non-paradigmatic. Following the historical inquiry, I shall develop a distinction between what I call *original* and *derivative* examples, a distinction that emerges directly out of a certain historical story. This distinction, differing from the traditional one between paradigm and borderline examples, rests upon a relation of conceptual dependency: briefly, a derivative example is conceptually dependent upon an original example, but not vice versa.[16]

III

This model of open concepts has theoretical effect upon our understanding not only of concepts but also of the objects that fall under them. One particular way to see this effect requires us to distinguish between identity conditions and ideals. This distinction serves to mediate the move from describing objects per se, to describing regulative concepts and the projected existence of objects. To introduce the distinction consider a passage from the *Philosophical Investigations*, which begins with Wittgenstein recalling the emphasis F. P. Ramsey once placed on the idea that logic was a 'normative science'. Wittgenstein took this idea to be closely related to another, that in philosophy 'we often compare the use of words with games and calculi which have fixed rules, but cannot say that someone who is using language must be playing such a game'. But, he continued,

if you say that our languages only approximate to such calculi you are standing on the very brink of misunderstanding. For then it may look as if what we are talking about were an ideal language. As if our logic were, so to speak, a logic

[16] Note that my theory has much in common with the institutional theory of art (cf. Arthur Danto's *The Transfiguration of the Commonplace: A Philosophy of Art* (Cambridge, Mass., 1981), and George Dickie's *The Art Circle: A Theory of Art* (New York, 1984)). But there is a difference. I have tried to avoid the tendency, sometimes visible in other institutional theories, to offer a comprehensive account of all the arts. Emphasizing generality and sameness at the expense of historical particularity and difference limits, in my view, the theory's full potential.

for a vacuum.—Whereas logic does not treat of language . . . in the sense in which a natural science treats of a natural phenomenon, the most that can be said is that we construct ideal languages. But the word 'ideal' is liable to mislead, for it sounds as if these languages were better, more perfect, than our everyday language, and as if it took the logician to show people at last what a proper sentence looked like.[17]

Wittgenstein was attacking two related ideas: that language is meaningful only if it consists in fixed (logical) rules, and only if it approximates to an ideal language (where an ideal language is understood as consisting of fixed and exact rules). Ideal languages, he argued, can be constructed as languages, but they do not and need not form part of all languages. An ideal language does not,.in other words, necessarily form a part of the natural languages we use. The idea that revealing the perfect logical form of language explains the meaning of the words or propositions in that language is therefore mistaken.

I believe ideals can play a role in our 'informal' cultural practices, though what I mean by 'ideal' is somewhat askew to Wittgenstein's meaning. Ideals are what we strive towards within our practices. Following Wittgenstein, however, these ideals can neither be explained independently of the relevant practice nor be articulated in terms of fixed rules. This can best be explained by extending an argument begun in the previous chapter against the search for identity conditions.

Recall Goodman's conditions of accurate notation and perfect compliance. These (like Levinson's conditions) were or can be articulated in terms of identity conditions. ('Something is X iff . . . ' or 'X is a necessary and/or sufficient condition for something to be the kind of thing it is.') Traditionally, these conditions have been conceived to be the sort of things that demand to be *met* or *satisfied*. For something to have the identity it does it must meet certain conditions. The 'must' here is of a Leibnizian logical sort, as is the notion of identity with which we have been dealing.

In the previous chapter, I suggested that the problem with identity conditions resided at the intersection between theoretical and empirical considerations. Only at this intersection did we see confusion over what could count as an identity condition and over what kind of justification could be offered for the positing of one. The problem resided in the incompatibility between the theoretical demands associated with identity conditions and the phenomena to be accounted for.

[17] § 81.

Most if not all identity conditions for works and performances are, in my view, mis-translations of ideals that exist within classical music practice. In all translation something is lost. In this case most of the empirical and ideological content is drawn off, leaving what we have seen to be little more than generic formulations. Consider how many types of action could be identified on the basis of that non-extensional compound condition offered earlier for something to be a performance. Something is a move in a tennis game if and only if the player intends it to be such, it is recognized as such, and it accords sufficiently with the rules of the game.

The argument against translation requires us to stop exclusively seeking identity conditions for objects and to describe on the level of concepts the sense in which works require of their performances perfect compliance with scores. Perfect compliance is an ideal we strive towards in the performance practice of classical music. It is an ideal of paramount evaluative and aesthetic importance that performers strive to produce perfect expressions of works. This ideal is an ideal because of certain aesthetic beliefs about what musical works are and what their performances should be like. Its existence is founded upon a complex aesthetic theory underlying the conceptual and institutionalized structure of classical music practice.[18]

What we understand today to be perfect compliance has not always been an ideal and might not be in the future. Actually it is quite peculiar and rather unique. It has characterized classical music practice only for the last 200 years. It is also not universal in the world of music. In fact, it is significantly this ideal that serves to distinguish the practice of producing performances of classical musical works from the performance practices associated with other kinds of music. Whereas in classical music performances we strive towards maximal compliance with a fully specifying score, in traditional jazz improvisations, where very different notions of compliance operate, musicians seek the limits of minimal compliance to tunes or themes. In most jazz, extemporization is the norm, and it is just this feature that forecloses the possibility of our speaking comfortably of one and the

---

[18] Cf. David Carrier's discussion of performance practice in his 'Interpreting Musical Performances', *Monist*, 66 (1983), 202–12. He speaks of the risk-taking involved in producing performances and of the way in which performers try to reduce the distance between the performance and the work itself.

same work (rather than of a tune, theme, or song) simply being instantiated in different performances.[19]

To speak of ideals is not to speak in a disguised form of identity conditions. Operating within the confines of a search for identity conditions, we are faced with a seemingly intractable tension between 'chemically pure' theory and the indeterminacies of practice. Operating on a conception of ideals we move towards a new tension. This one turns upon a dichotomy inherent in all cultural practices, that between the ideals supporting a practice and the limitations of human action. This tension is a positive one. It tells us that we cannot stipulate that every performance *be* perfectly compliant with the score, only that it is an *ideal* that each performance be such. Recognizing something to be an ideal means that it is rarely if ever perfectly realized, but this does not undermine its existence and force in any way.

If perfect compliance is an ideal of musical practice, it cannot also be a *prerequisite* for performance individuation. To see it as such would be to ask too much of the practice; it would be to over-determine it. In other words, it would be a mistake to specify perfect compliance as an identity condition for a performance of a musical work.[20]

Focusing on ideals paves the way to an understanding of performance individuation. A similar emphasis contributes also to our understanding of the purpose of a score. Goodman argued that for a score to serve the function of identifying a work in a chain of score-copies and performances, the notation must meet strict requirements. I prefer a description

[19] Wynton Marsalis expresses the point in rather different terms: 'Concert musicians are artisans—Jazz musicians are artists. . . . With Bach or Haydn, you know what you're playing is worth hearing, and the best thing you can do is not mess it up. In jazz, you have to have something worth saying and then know how to say it.' (Quoted in B. Buschel, 'Angry Young Man with a Horn', *Gentleman's Quarterly*, 57 (1987), 195.)

[20] Someone might suggest that no identity condition is a translation of an ideal, let alone most or even some of them. The perfect compliance condition might have nothing to do with the ideal of perfect compliance operative in practice; the former is to be understood as a theoretical condition having only to do with issues of identity, etc. If so, my mention of translation is inadequate. Still, I am using the idea of translation only to highlight a difference between two philosophical languages. Perhaps, moreover, identity conditions are no more than theoretical idealizations derived from observation of the relevant phenomena, and any theory articulated in terms of such conditions should be judged on the basis of its explanatory power. This would concur with Aristotle's conception of generalization: generalizations and the theories within which they appear are modified according to the degree to which they explain the phenomena. Were one to take this position, identity conditions might begin to coincide with what I am calling the ideals of a practice. But, again, what I am arguing against here is the theoretical weight accorded to identity conditions as they have been used in the literature, and that weight is stronger than that described in Aristotelian terms above.

which admits the existence first of an ideal, an ideal that says that the score should be as accurate as possible, and second of a practice, a practice in which the ideal is never quite lived up to, but none the less functions.

If it is an ideal, we cannot conclude that perfect notation is a prerequisite for the functioning of the score, just as we cannot conclude that perfect compliance is a prerequisite for performance production. And just as we can then ask how performance individuation functions without this prerequisite, we can ask how scores manage to serve their function despite the inevitable failure of notational precision. As in the case of performances, an adequate story appeals to a complex, institutional understanding of music.

A final difference between identity conditions and ideals is that the latter, unlike the former, are action-guiding. When we act in accordance with an ideal, we act in a domain of normativity. We are guided by certain beliefs and we develop appropriate skills. All of this comes to be reflected in the institutionally generated expectations that are bound up, for example in the musical world, with our producing and recognizing events as performances of musical works.

In sum, ideals and identity conditions carry different theoretical weight: the latter demand to be met, the former do not. That something functions as an ideal is insufficient evidence to guarantee its functioning as an identity condition. Ideals cannot easily be translated into conditions. To move away from conditions and towards ideals is to move towards a discussion of the conceptual understanding implicit in musical practice, since talking about ideals forces us to look at the historico-conceptual foundations of practice in a way the traditional search for identity conditions for musical objects does not.

Unlike identity conditions, ideals cannot exist in 'a logical vacuum'. On the contrary, having a basis in a logical, coherent, or fully rational theory is not a pre-condition for ideals to exist in a practice. Ideology (in its least derogatory sense) and its expression in practice suffice. Many ideals rest on dubious theoretical foundations, but they still function as ideals. Of course, the better the ideology, the better the ideal—whatever 'better' means.

IV

To see musical activities working alongside the ideals of a practice matches the idea that some concepts are regulative. The notion of a regulative concept was used with significance by Kant and was brought into the modern discipline of aesthetics notably by Wellek and Warren. These latter theorists suggested that artistic genres function regulatively. They could also have seen regulative concepts to function in particular as well as in general terms. Instead of asking what it means to say that the concept of a work of art is regulative, they could have proposed that each work itself is a regulative concept. When one speaks of works in this way it might however be better to speak of them as norm kinds, or as structures or sets of norms, which in fact Wellek and Warren do.[21] This view of individual works is not without recent proponents. In Chapter 1, brief references were made to Wolterstorff's, Tormey's, and Walton's normative views. My view is different from all of theirs, however, because it keeps normativity and regulation on a general level. I suspend (without loss) consideration of the former view and speak not of each work, but of the concept of a work, as a regulative concept.

Regulative concepts differ from constitutive ones: the latter constitute the fabric of a practice; they provide the rules of the game. New constitutive rules signal either a new game or a new version of an old game. Regulative concepts guide the practice externally by indicating the point of following the constitutive rules. In moral practices, certain constitutive rules are provided to indicate what one should and should not do. The point of following these rules is founded upon our grasp of concepts such as those of freedom, justice, and responsibility. The latter do not make up the structure of the practice; rather, in their interrelations, they determine what the structure should be like.[22]

In their normative function, regulative concepts determine, stabilize, and order the structure of practices. Within classical music practice we compose works, produce performances of works, appreciate,

---

[21] See Kant's *Critique of Pure Reason*, tr. N. Kemp Smith (New York, 1929), Transcendental Dialectic, III. 8. 449, 'The Regulative Employment of the Idea of Pure Reason'; also R. Wellek and A. Warren, *Theory of Literature* (New York/London, 1977), 261, 265, and 150, who write: 'The real poem must be conceived as a structure of norms, realized only partially in the actual experience of its many readers. Every single experience . . . is only an attempt—more or less successful and complete—to grasp this set of norms and standards' (150). They then point to the difference between regulative concepts and individual works: 'Though the genre will appear in the history exemplified in the individual works, it will not be described by all traits of these individual works.' Finally, they point out that 'we must conceive of genre as a "regulative" concept, some underlying pattern, a convention which is real, i.e. effective because it actually moulds the writing of concrete works' (261-2).

[22] I have benefited from J. Rawls's 'Two Concepts of Rules', *Philosophical Review*, 64 (1955), 3–32, and Tormey's 'Indeterminacy and Identity in Art', 210.

analyse, and evaluate works. To do this successfully we need a particular kind of general understanding. Every time we talk about individual musical works we apply this general understanding to the specific cases. This understanding focuses upon one or more regulative concepts.

The force of regulative concepts is one of guidance through *phronesis* and example, rather than dictation by explicit or formalized rules abstracted from practice. As Polanyi argued,

Maxims are rules, the correct application of which is part of the art which they govern. The true maxims of golfing or of poetry increase our insight into golfing or poetry and may even give valuable guidance to golfers and poets; but these maxims would constantly condemn themselves to absurdity if they tried to replace the golfer's skill or the poet's art. Maxims cannot be understood, still less applied by anyone not already possessing a good practical knowledge of the art.[23]

Regulative concepts, like the ideals with which they are connected, function in a practice if participants act in a learned way. All activities are learned activities embedded in domains of normativity.

A regulative concept determines the normative content of subsidiary concepts as it does the content of associated ideals. Dialectically, the subsidiary concepts and ideals determine the normative character of the regulative concept. There is no temporal or logical priority in either determination. Rather, a set of concepts emerge in relation to one another, one or more of which may achieve a position of central regulative force.

The concept of a musical work, for example, emerged in line with the development of numerous other concepts, some of which are subsidiary—performance-of-a-week, score, and composer—some of which are oppositional—improvisation and transcription. It also emerged alongside the rise of ideals of accurate notation and perfect compliance. In this process, the work-concept achieved the most central position.

But what is the content of the work-concept or of any other regulative concept? Though precisely named perhaps, the content of regulative concepts is usually quite elusive. Hence, it is normal to understand them vicariously. They are understood by reference to the

---

[23] *Personal Knowledge* (London, 1962), 31. On the concept of rules and roles and how they are connected to forms of appropriate behaviour, see R. Sennett, *The Fall of Public Man* (New York, 1977), ch. 2.

ideals of a practice, to subsidiary constitutive concepts, and even more concretely to a set of beliefs and values about the status and nature of the objects and activities of a practice.

Correspondingly, when we act in accordance with a regulative concept we act vicariously. We act on our beliefs and values articulated in terms of constitutive rules. When we act we do not always explicitly think of the 'higher' regulative concepts with which these rules are associated. The way practice works is not necessarily identical to how people think it works in their day to day activities. Persons tend to think globally only when encouraged or forced to do so. Furthermore, when we act in accordance with a set of rules, beliefs, and associated ideals, we recognize their interrelations. Like a set of rules for a game, when we know how to act in relation to one belief, rule, or idea, we usually know how to act in relation to the others.

Regulative concepts are delimiting. They indirectly suggest to the participants of a practice that only certain beliefs and values are to be held and only certain kinds of actions are to be undertaken. In this sense, regulative concepts are structuring mechanisms that sanction particular thoughts, actions, and rules as being appropriate. Thus, for example, performing a work involves employing the appropriate regulative concept(s). One shows one's knowledge and understanding of these concepts when one, for example, complies with a score, plays these notes and not others, plays in such a way as to indicate respect for the genre musically and historically conceived.[24]

Regulative concepts function in a stable manner as they come to be entrenched within a given practice. How is this possible? Following a Kantian line of thought, these concepts function stably because they are treated as if they were givens and not 'merely' as concepts that have artificially emerged and crystallized within a practice. That they are treated as givens does not mean they are so. They become anchored in a practice through a kind of fictional or suppositional permanence. In this way they are seen to provide the ultimate grounding, the externalized and thereby transcendent principle of ordering, the externalized point of reference, for the practice.

---

[24] Foucault describes the situation well when he argues that a given ideal, concept or principle is seen to determine or delimit through its regulative capacity a domain of behaviour, objects, rules, tools, and discursive practices. Through this capacity it sets up the criteria according to which these constitutive parts of the domain are judged to be relevant, central, or marginal. See his *The Archaeology of Knowledge and the Discourse on Language*, tr. A. M. Sheridan Smith (New York, 1972), 61.

Certain musical and artistic concepts are postulated as given for the same kind of regulative rather than constitutive reasons that other kinds of concepts, such as those of God and of human freedom, are postulated. The latter are postulated as if they were transcendental or given, to justify our faith in moral judgement and responsibility. They are postulated for the same kinds of reasons Mill famously invoked as according an ultimate sanction to the principle of utility. Similarly, musical concepts are accorded this status to stabilize a given kind of musical practice. Even in artistic practices some things do not (and ought not to) happen. We do not, for example, improvise while performing Beethoven's Violin Concerto in D 'seriously' in a concert hall.[25]

The stability of our practices requires that their central concepts retain their regulative force. This is achieved, I have just suggested, by agents suspending the belief that concepts artificially emerge, that their identity and stability depend on the demands and theoretical underpinnings of the practices. The concepts are artificial, but not in a derogatory sense. We simply think about them as if they were absolute (often with unfortunate consequences) to avoid the challenge of a threatening opposition or the challenge of relativists. Another way to put the point is to say that from within the practice, regulative concepts are seen to be self-legitimating. Externally, they and the practice they regulate require a different kind of legitimation. What form does this external legitimation take? The answer is complex, and rests upon the idea that concepts emerge within a long and complicated process. The answer requires that we know many things: how, for example, concepts emerge and acquire their regulative status; how concepts are used during their emergence, at the point when they are fully emerged, and then after that point.

---

[25] Mill, *Utilitarianism, On Liberty and Considerations of Representative Government*, ed. H. B. Acton (London, 1972), 27 ff. The analogy between musical and moral concepts, resting on the shared ability to produce order and regularity in the respective practices, is captured well in the words of Ssu-ma Chien: 'The sacrifices and music, the rites and the laws have a single aim; it is through them that the hearts of the people are united, and it is from them that the method of good government arises' (quoted by Attali, *Noise*, 29). Aristotle also wrote to this point: 'As is the steersman in the ship, . . . the laws of the city, the general in the army, so is God in the Universe; . . . he moves and revolves all things, where and how he will, in different forms and natures; just as the law of a city, fixed and immutable in the minds of those who are under it, orders all the life of a state' (*De Mundo*, ch. 6: 399a, 400b (*The Works of Aristotle* ed. W. D. Ross (Oxford, 1931), iii)).

Finally, any description of regulative concepts is complicated by the fact that such concepts are not necessarily associated with one set of beliefs or values. Over time we may act in accordance with a concept in many different ways as the needs of a practice change. To act in accordance with a concept never requires, though it might contingently end up being the case, that we hold those beliefs and values originally associated with it. It is through continuity of function, however, that we assess the situation to be one in which we are using the same concept over a period of time. This view matches the description of open concepts given above.

V

What now are we to say about the existence of the individual works that fall under the regulative work-concept? In other words, do musical works exist? In its regulative capacity the work-concept suggests to us, because of some quite peculiar aesthetic and musical reasons offered at a particular time, that we should talk of each individual musical work as if it were an object, as if it were a construction that existed over and above its performances and score. In a projectivist view, indicated by the 'as if' clause, works do not exist other than in projected form; what exists is the regulative work-concept. However, insofar as this concept functionally involves projections or hypostatizations—for each work composed we project into it 'object' existence—the resultant objects are accorded projective or fictional existence. A fictional object was succinctly described by Bentham as 'an entity which, though by the grammatical form of the discourse employed in speaking about it, existence be ascribed, yet in truth and reality existence is not meant to be ascribed'.[26]

The most convincing way to describe the motivation for our projecting the existence specifically of musical works is via a description of the quite peculiar empirical and historical rationale that founds the belief and then the hypostatization that there are musical works. Everything about musical practice of a classical sort suggests that composers produce works and not just performances and scores. All our activities of performance, criticism, and evaluation rest on this supposition. But to what does the supposition amount?

---

[26] J. Bentham, 'Of Fictitious Entities', in C. K. Ogden (ed.), *Bentham's Theory of Fictions* (New York, 1932), 12.

Given the projectivism thesis, the historical inquiry justifiably commences on the level of concepts and not objects. The inquiry moves away from describing the status of the individual works themselves. Is there a problem here? Consider the following premisses: (i) the practice of classical music functions in line with a complex theory about what the practice is and should be like. (ii) Within that theory, musical works are posited as existants. (iii) Successful positing of works confirms the effective functioning of the regulative concept of a work within this practice. Given (i)–(iii), does it follow that the subsequent historical account of the work-concept reflects a theory about musical practice but not the practice itself?

A positive answer is generated when it is presupposed either that theory and practice are structured and determined independently of one another, or that a practice has a 'real', underlying structure existing independently of a theory of it. But I do not presuppose either statement especially if they presuppose something other than the fact that the distinction between theory and practice is formed as a convenient, intellectual abstraction. While admitting of many shades and degrees, the distinction enables one to examine a single (unnamed) thing from a multitude of different perspectives. Practice, however, ultimately remains theorized, as theory ultimately remains practised. Concepts and projections have as much pragmatic import and expression as they do theoretical. To generate too much of a rift between theory and practice is to ignore the import of premisses (i) and (iii).[27]

VI

I have suggested many times that the work-concept emerged. What does that suggestion entail? Emergence depends upon the development and crystallization of numerous moments ranging from the introduction of more and less specific theories that found new beliefs, values, concepts, and distinctions; to new rules, laws, and normative principles; to new tools and institutional buildings; to new activities and modes of behaviour.

Emergence is not a process having its source in an original (essential) seed. 'It is one thing to see the interconnectedness of things,' Ernst Gombrich writes, 'another to postulate that all aspects of a culture can

---

[27] I have benefited from R. Geuss's discussion in *The Idea of a Critical Theory: Habermas and the Frankfurt School* (Cambridge, 1981), 55 ff.

be traced back to one key cause of which they are all manifestations.'[28] Emergence is not a pre-determined process showing the inevitability or predictability of the rise of a given concept. It is rather a contingent, retroactively discovered, bonding and roping process.

A concept emerges out of the roping together or fusion of the moments just described. Only when it has emerged can we retroactively discover its original threads. Emergence is not a process of creation *ex nihilo*, but a slow process occurring through the development of a practice, through the fostering of new theories and much more. A concept's emergence is not, furthermore, an event divorceable from its past or from the history of the practice within which it functions. The idea is to see emergence as part of the history of the practice itself.

Prior to the point at which we would say a concept has emerged, it might be that many if not all the threads of what becomes the content of the concept already exist. As yet however they have not meshed together in the appropriate way to admit the concept's regulative function, if the concept has such a function. This phenomenon helps explain why, when movements transform themselves one into another, the new appears as much continuous as it does discontinuous with the old.

Thus, prior to 1800 there were functioning concepts of composition, performance, and notation in musical practice, just as there were after that time. This is the continuity. The discontinuity lies in the fact that their significance, and the conceptual relations in which these concepts stood to one another, differed across the two time periods. Later we shall have reason to conclude that the work-concept had a regulative

---

[28] *In Search of Cultural History*, 30. My conception of emergence contrasts with the 'Whig Conception of History'. It is more in line with Arendt's conception, since, in her investigation of political concepts, she differentiates her approach from straightforward causal history, as Foucault, with similar intent, distinguishes his approach from a casual history of ideas. We all share the belief that to describe the emergence of concepts is not to describe causal factors, if that would entail describing the underlying necessity of the emergence. 'The elements of totalitarianism form its origins,' Arendt thus writes, 'if by origins we do not understand "causes". Causality, i.e., the factor of determination of a process of events in which always one event causes another and can be explained by another is probably an altogether alien and falsifying category in the realm of historical and political sciences. Elements by themselves probably never cause anything. They become origins of events if and when they crystallize into fixed and definite forms. Then, and only then, can we trace their history backwards. The event illuminates its own past, but it can never be deduced from it.' ('The Nature of Totalitarianism' (1954), unpub.; Library of Congress, quoted in E. Young-Bruehl's *Hannah Arendt: For the Love of the World* (New Haven, Conn., 1982), 203.)

function in the later, but not in the earlier, period, despite the presence of continuity in both theory and practice.

The emergence of a regulative concept covers the period when the concept is in gestation. During this time, if the concept functions at all, it is by intimidation and without stable meaning. During this time the content of the concept crystallizes. After this period, the concept explicitly functions in its regulative capacity. Now it functions in an entrenched, stable, and accepted manner. From a practical point of view, it is usually during this time that the concept sinks into opacity; its existence is taken so much for granted that we find it difficult to think of the practice without it. To adapt Hegel's teaching: what increasingly becomes familiar to us increasingly becomes unknown to us just because of our feeling of familiarity.[29]

From a theoretical point of view, it is only when the concept becomes familiar to us that we can begin critically to comprehend its meaning and function. Not only can we begin to trace the whole range of its possibly changing regulative meanings and functions, but we can also retrospectively trace its unstable meanings during the period of its emergence. Taken together, these tasks match the intention to trace the history of the musical work as an open concept in the different stages of its life.

But how do we identify the initial, unstable, and then the regulative uses of a concept? Dictionaries help. They indicate known first uses, and they tell us by the absence of terms if terms were not used at least in recorded speech at a given time. Still, they are inadequate to tell us whether concepts were present at a given time or how they were functioning, if they were functioning at all. For this we need to look at the theories, beliefs, laws, and activities existing during that time.

The latter sort of inquiry might reveal that participants of a practice could not have been using a given concept regulatively because the relevant conceptual use was overpowered and excluded by the presence of alternative regulative concepts. That is just the sort of information we need, for with it, we can then trace the changes occurring in these

---

[29] Hegel's *Phenomenology of Spirit*, tr. A. Miller and J. N. Findlay (Oxford, 1977), 18: 'The familiar, just because it is familiar, is not cognitively understood. . . . Subject and Object, God, Nature . . . are uncritically taken for granted as familiar, established as valid, and made into fixed points for starting and stopping.' Cf. R. E. Palmer's analogous description of tradition: 'Tradition . . . is something in which we stand and through which we exist. For the most part it is so transparent a medium that it is invisible to us . . . as invisible as water to a fish' (*Hermeneutics* (Evanston, Ill., 1969), 177).

theories and activities to locate an approximate date when the concept began to function regulatively. Finding a 'rough' date is satisfactory because conceptual change, like the change in practices, has no sharply defined beginning or end.

Just as precise dates are not forthcoming in our inquiry, so neither are definitions of a typical philosophical sort. In the literature, it has been argued that our philosophical interest in concepts does not have to focus on matters of philosophical definition, as it so often has done in the tradition. The alternative argument asks us to consider what it means to act knowledgeably, correctly, and appropriately within a given practice. What must be done to bring about something (an event or an object) that is appropriate in given circumstances? What must be done to bring about, say, a performance of a given musical work? Given a Rylean quest for knowledge of this kind, one could claim that to understand the employment of a concept no philosophical definition is required. The knowledge of when and how we use a concept requires something different from the possession of a definition of this sort.

William Kennick once presented this kind of argument when, with his famous warehouse test, he suggested that persons could pick out reasonably successfully those objects they thought were works of art without employing a definition. He claimed, in fact, that were one to have a definition in mind, say 'art is significant form', one would actually find it harder to distinguish artworks from non-artworks. One might immediately want to retort here that if subjects were ever asked to give reasons for choosing these rather than those objects as examples of artworks, they would appeal to a definition. With regard both to classificatory needs and performance practices, some definitional understanding is always presupposed. This counter-claim is telling. However, Kennick's argument had a provision. He was arguing against philosophical definition and not against definitions of other sorts.[30]

Kennick was not referring to definitions with which we (perhaps unconsciously) operate in everyday activities. He did not deny that in everyday activities our actions are guided by something like a definition, articulated perhaps in terms of paradigm examples, prototypes, or personally selected features. Any of these can be used as a standard to classify examples. 'Working' definitions are indispensable in fact; they help determine appropriate forms of performance and evaluative

[30] W. E. Kennick, 'Does Traditional Aesthetics Rest on a Mistake?', *Mind*, 67 (1958), 327.

behaviour. Much of our appreciation of something turns upon our recognition that it is a good example of its kind. In most cases, one must know what kind of thing one is looking at in order to respond to it appropriately.

Speaking about working definitions accords with the theory of open concepts. (Kennick would have been sympathetic to this.) As with that theory, the description of a 'non-philosophical sort' of definition has more relevance to the kinds of issues of classification and action to which a description of practice gives rise. So what are we to do now with definitions of the 'philosophical' sort? Should we dispense with them? Not altogether. Theories and definitions of a philosophical sort play a major role in the development of practices. How a definition (say 'art is significant form') is transformed into a set of practical and often personal beliefs and values about art is as subtle and complex an issue as how philosophical theory and practice coincide with one another.

The historical inquiry will point to many aspects both of this transformation and of this coincidence. None the less, unlike other philosophical treatises in aesthetics, it will not yield a definition that is claimed to be a 'philosophical' definition of 'musical work'.

VII

We are finally in a position to examine the philosophical content of the central claim, that the work-concept began to regulate musical practice at the end of the eighteenth century. One way to do this is via prima-facie objections that might be put to the claim. Continuing in philosophical mode, one might, for example, challenge the notion of workhood employed to sustain the whole argument.

Recall that I began this essay with a description of musical works which I said was given in pre-critical terms. Works were described as existing as public and permanent artefacts, created by composers, and constituted by structures usually of sounds, dynamics, rhythms, and timbres. And so on. Ignoring this description, someone might suggest (and this has been suggested to me many times) that musical workhood covers a spectrum of cases ranging from the most neutral and general to the most contentful and specific, from the ideologically free to the ideologically specific. The description of works I have given falls at the latter end—the specific and the ideological.

One might choose to use the notion of workhood to pick out any musical unit—any string of notes or sounds, even if that meant concluding that the sound of the wind in the trees or a popular song are works as much as Beethoven's concertos are works. Not wishing to go that far, one might seek identity conditions of the sort mentioned in previous chapters. Perhaps a work is any combination of sounds strung together in a deliberate operation of initiation. Where one would now exclude sounds of the wind from the class of works, one would still include popular songs and many other formations. To exclude further undesirable items, one would just increase the number of conditions. One might emphasize the aesthetic function of performing certain combinations of sounds.

Not willing to accept any such conditions, on the grounds that their use detracts from the neutrality of the concept, one could claim that any musical unit, however that is understood, can count as falling under the work-concept. Why not just acknowledge that the work-concept is a category under which bits of music fall, however 'music' is described? This, or so it might be said, is really what is meant by claiming in neutral terms that a musical work is any (musical) combination of (musical) elements.

Though this claim fails to reflect the use of the concept in the history of musical thought and practice, where would it have got us? Only as far as the circular and vacuous knowledge that a musical work is any musical combination of musical sounds (put together by a musician?).[31] Were one now to return hurriedly back to the set of identity conditions to generate a non-circular definition, one would find oneself hard-pressed to tally its use with a description of a neutral and general concept. For the conditions do not actually have any content until clothed within a real context of use, and that use points to specificity rather than generality. To speak of a compositional condition or that of deliberate operation is to presuppose, unless specified otherwise, a whole understanding of what it means to compose music; *a fortiori*, of what it means to notate and perform music and of what it means to listen to music aesthetically. In fact, when that understanding is made fully explicit, it does not motivate one to conclude that the work-concept is near to being neutral or ideologically free; quite the reverse.

---

[31] Cf. Cone's discussion of Tzara's circular definition in 'One Hundred Metronomes', 446.

The 'pre-critical' description of musical works appears pre-critical only because it is so familiar. It matches how those 'in the classical mainstream' think about works. But critically, the description has its roots in a peculiarly romantic conception of composition, performance, notation, and reception, a conception that was formed alongside the emergence of music as an autonomous fine art.

Now one might worry that I've gone too far, that this description of workhood weds it far too closely to one historical moment. Surely, to say there is no neutral work-concept does not mean the concept must be glued to so particular an historical understanding. Surely the more plausible view is that the work-concept has functioned with varying meanings over time. To see how justified these worries are, let us investigate precisely to what the central claim is committed.

The claim is that given certain changes in the late eighteenth century, persons who thought, spoke about, or produced music were able for the first time to comprehend and treat the activity of producing music as one primarily involving the composition and performance of works. The work-concept at this point found its regulative role. This claim is not committed to the supposition that the work-concept has, since this time, retained its original foundation in the sense that it has come to take on no further meaning. Nor does it imply that composers producing music in centuries prior to the nineteenth were not producing works. Thus, despite the story of its emergence into a regulative concept in the late eighteenth century, the use of the work-concept is not confined to products only of this and later periods.

Consider that, even if circumstances do not allow for conscious or explicit use of a concept, what is produced under these circumstances may still fall under the concept. Maybe Bach produced works, even though he explicitly thought about music in different conceptual terms. That may be so, but it is not so in any straightforward sense. Can a concept have, in fact, a form of existence, namely implicit existence, over and above explicit existence? One can view implicit existence logically. A concept implicitly exists before its explicit emergence if and only if it was logically possible at a given time for it to have emerged explicitly at some other time. Or, focusing on the contrast between conscious and unconscious conceptual use, one may claim that to say a concept functions consciously at one time is to assume that it functions unconsciously at another (either earlier or later). Neither clarification has to entail the realist assumption that the relevant concept always existed (either implicitly or explicitly) even though we might not always have known it. Such claims can incorporate temporal constraints.

Ignoring the impending logical complexities, I am interested above all in resisting the inclination to say that the work-concept must always have functioned in some manner. The work-concept is not a necessary category within musical production. To make that conclusion harder to draw, I will give the claim about implicit existence a further nuance.

What is in question is whether musicians before 1800 (or at any time), though conditions forbade overt expression, none the less were thinking about musical production in terms of works. There are historical reasons for rejecting this conclusion. There is also a philosophical way to avoid it. Thus, there is an epistemological sense in which implicit function depends upon a concept's having functioned explicitly *first*. Only with its explicit function realized can we in hindsight see the concept as functioning implicitly. Prior to its explicit emergence, there is no evidence to suggest that persons were *really* (whatever that means) thinking about something in conceptual terms distinct from those indicated by their expressed thought and behaviour. Certainly there are persons who led the way forward, but even their prophetic role was recognized as such only after the event.

This epistemological claim presupposes that certain if not all kinds of meaning and truth are dependent upon the existence of particular conceptual schemes. That presupposition allows one to affirm that, given that we have an explicit concept of a work, Bach composed works. If the concept had never acquired its explicit, regulative function within musical practice (or indeed within any other relevant or related practice), we would probably still speak of Bach's music in terms not only more familiar to Bach himself, but also still evident in other existing musical practices not regulated by the work-concept.

The claim can be made more precise. We need to distinguish two potentially distinct claims: first, prior to 1800, the work-concept existed implicitly within musical practice; second, prior to 1800, the work-concept did not regulate practice. They might be different claims but they can usefully function together. First, prior to 1800, musical activity was not structured by the work-concept. What was produced before this time was seen to fall under concepts other than that of a work. If musicians used the term 'work' (or a synonym) at all, their uses did not reflect a regulative interest in the production of works. It is in this sense that I earlier suggested that Bach did not think about his activity predominantly or intentionally in terms of works; his activity was otherwise regulated.

Yet, second, Bach did compose works. Given the epistemological dependency between implicit and explicit existence, we can speak of Bach as having produced works in the following way. This way depends upon our importing a conceptual understanding given to us when the work-concept began to regulate practice. Just as a piece of pottery or a pile of bricks can come to be thought of as, or transfigured into, a work of art through the importation of the relevant concepts, so, since about 1800, it has been the rule to speak of early music anachronistically; to retroactively impose upon this music concepts developed at a later point in the history of music. Implicit existence has become here essentially a matter of retroactive attribution.

Now we can make sense of the basic argument lying behind my central claim that prior to 1800 (or thereabouts), musicians did not function under the regulation of the work-concept. To be sure, they functioned with concepts of opera, cantata, sonata, and symphony, but that does not mean they were producing works. It was only later when the production of music began to be conceived along work-based principles that early operas, cantatas, symphonies, and sonatas acquired their status as different kinds of musical work. And this is why we can meaningfully say, nowadays, that Bach composed musical works.

VIII

Much of the information I have provided in this chapter can be used as well to sort out a remaining general issue. But to sort the issue out satisfactorily, I need to introduce a new historical dimension into the argument. That 1800, or thereabouts, should be the point at which the work-concept became regulative might be judged controversial in the light of evidence often brought forward to show the beginnings of this concept's existence in the early sixteenth century. It is often claimed, in other words, that the concept was in use in centuries prior to the nineteenth.

Nearly all theorists (mostly German musicologists) who have looked at the emergence of the work-concept have located two historical moments as worthy of note: c.1527 (sometimes 1537) and 1800. Regarding the earlier date, it has often been claimed that the origin of the work-concept is to be found in Nicolai Listenius's book on music education in Reformation Germany. It is worth quoting Listenius in full.

> Music is the science of producing melodious sounds correctly and well. To produce such sounds well is to produce a song subject to some fixed rule and measure properly, through its tones and notes. And it is threefold. *Theorike* [*sic*, for *Theoretike*], *Praktike*, *Poietike*.
>
> Theoretical [Music] is that which is concerned solely with the contemplation of natural capacity and the understanding of the subject [*rei cognitione*]. Its goal is to know. Hence the Theoretical Musician who has learned this art, truly content in this alone, presents no example of it by performance.
>
> Practical [Music is that] which does not only lie hidden in the inner sanctum of natural capacity, but issues forth in an *opus* [action, performance, work, labour], though no *opus* remains after the performance. The goal of Practical Music is performance. Hence the Practical Musician [is one] who teaches others something more than an understanding of the art and trains himself in it around the performance of any *opus*.
>
> Poetic [Music is that] which is not content with either an understanding of the subject or with practice [*exercitio*] alone, but rather leaves some *opus* behind after the labour, as when music or a musical song is written by someone, whose goal is a complete and accomplished *opus*. For it consists in making or constructing, that is, in such labour that even after itself, when the artificer is dead, leaves a perfect and absolute *opus* [*opus perfectum et absolutum*].[32]

The phrase *opus absolutum et perfectum* has been interpreted to indicate the presence of the work-concept. Theorists have then argued that other developments in the sixteenth century, say, in compositional techniques, in the production of operas, and in the publication of music, further demonstrate the concept's presence. To my mind, the argument does not stand up to critical scrutiny. The appeal to Listenius is difficult to sustain.

Listenius was neither Robinson Crusoe nor an exceptional genius. Like others, he was influenced by intellectual ancestors. He was influenced by the writings of Antiquity, especially Aristotle's *Poetics*, and particularly Aristotle's use of the tripartite distinction applied to

---

[32] Nicolai Listenius, *Musica: Ab authore denuo recognita multisque novis regulis et exemplis adaucta* (1549); facsimile, ed. G. Schünemann (Berlin, 1927). The crucial sentences on *musica poetica* are given in Latin as follows: 'Poetica, quae neque rei cognitione, neque solo exercitio contenta, sed aliquid post laborem relinquit operis, ueluti cum a quopiam Musica aut Musicum carmen conscribitur, cuius finis est opus consumatum & effectum. Consistit enim in faciendo siue fabricando, hoc est, in labore tali, qui post se etiam, artifice mortuo opus perfectum & absolutum relinquat.'

many spheres of human activity between knowledge (*episteme*), doing or activity (*energeia*), and making or producing (*ergon*). Listenius was not even the first to apply such a tripartite distinction to music; Boethius and Tinctoris plausibly preceded him.[33]

Connections do exist between Listenius's *opus absolutum et perfectum* and the work-concept. Yet they also exist between both the former and the ancient notions of art, work, and tragedy. Does this imply the origin of the work-concept is to be found in Aristotle or in ancient thought more generally? In a certain sense it does. But to identify the origin here would not be to tell the whole story about the musical work-concept. Three points follow. First, the work-concept is much related to Listenius's *opus absolutum et perfectum*, as it is to many ancient concepts. Second, even if it were true that Listenius gave crystal-clear expression to *musica poetica* (which he did not), this would not mean the work-concept originated with him. Third, looking for origins always threatens an infinite regression backwards through history. I shall treat the first two points together, the third separately.

Listenius's terms are not as clear as they have been thought to be. His description of *musica poetica* does not justify a proclamation made centuries later that 'Here's the evidence we've been looking for.' To begin, there is an Aristotelian reading of the controversial phrase that is just as plausible as the reading which equates *opus absolutum et perfectum* with the work-concept. The Aristotelian reading suggests that the phrase denotes a finished product which is the outcome of work or activity (*opus/labore*). One perfects one's work in the sense that one performs as best one can, with the goal of producing a perfect synthesis of skill and idea. An opus is the product of performance, not just the pre-existing idea that brings a performance about. A finished performance is as perfect and absolute as any activity of making that yields a separate, concrete, product. Both remain after the event as having been perfectly performed. An imperfect performance is 'dead'; unlike a perfect performance, it provides no normative rule for the performance of music thereafter.

Even if this reading were mistaken, that would not automatically entail the correctness of the other reading. Listenius does not explicitly indicate, for example, that an *opus perfectum et absolutum* would have existed as a completed creation *prior* to performance, or that, as a

---

[33] Cf. Seidel, *Werk und Werkbegriff*, 1–8, and Wiora, *Das musikalische Kunstwerk*, 17 for comparable and valuable discussions. I have found Seidel's discussion of Listenius extremely useful.

'musical work', its musical or sounding structure (rather than its words or text) would have been fully preserved in a score after a performance. He does not indicate, furthermore, that it would have been conceived as a uniquely and self-consciously created composition *repeatable* in its entirety in more than one performance. Lasting and being repeated do not necessarily mean the same thing. Listenius does not indicate, finally, that a work as a product existing over and above its performance would have been the primary point (*telos*) for musical activity. As far as I can tell, Listenius makes no further use either of the phrase *opus perfectum et absolutum* or the concept of *musica poetica* in his general text. *Musica poetica* seems to have been introduced to give emphasis to the idea that an understanding of making music—compositional principles—was as important as performing or theorizing about it. This turned out to be a particularly important theme in musical theory and practice from the sixteenth century on. Notwithstanding, one can make or compose music without thereby producing a work.

Of course, Listenius might have introduced or reflected an understanding of a work-concept prevalent in his time that preceded and gradually developed into our modern one. Whether that is true, and I think it might be for complicated reasons that shall be specified in Chapter 7, it has little bearing on my central claim. It would only have bearing on my claim if his use of the work-concept showed that it regulated musical practice. Evidence suggests that it did not.

Back now to the search for origins. One way to avoid infinite regression is to distinguish the search for origins from a description of a concept's emergence into a regulative concept. To see the point of this distinction, consider the second date accorded significance by theorists. As I said earlier, they have focused not only on the early sixteenth century but also on the years surrounding 1800.

Actually the latter date has received as much if not more attention than the former. While recognizing the 'origins' much earlier, many have ended up describing concepts associated with the work-concept that were explicitly articulated (as they also see it) at the end of the eighteenth century. They have looked at notions of autonomy, form, and product as these found expression especially in German romanticism. With this evidence in hand, they have then claimed that the work-concept, as Dahlhaus puts it, fused into consciousness in 1800 despite having originated at least two centuries earlier. Is it possible that their claim has ended up being the same as mine? Not quite, but nearly so.

We all agree that something happened around 1800. How we describe the event and what conclusions we draw from it differ in specific ways.

I have chosen to look for the emergence of a stable and well-founded use of the concept—what I have been calling its regulative use. Consequently, I place less weight than other theorists on the identification of original uses in early centuries of many *associated* concepts. There is a reason for this. In looking for the moment when musicians first used concepts of composition, performance, autonomy, repeatability, permanence, perfect compliance—concepts associated with the work-concept—one should not assume that any single one of these uses indicates the presence of a regulative work-concept. To assume this might lead one to ignore the important fact that, for a concept to function regulatively, many associated concepts have to function together and stand in the appropriate relations to one another in a particular way. Like other theorists, furthermore, I acknowledge that there were uses of terms prior to 1800 that came to be synonymous or nearly so with the term 'work'. 'Piece', 'composition', 'opus', are examples. But, again, unless the evidence can support it, one cannot assume that all these terms and uses of concepts indicate that musicians were thinking predominantly about music in terms of works.

I have chosen also to describe the emergence of the work-concept with its full aesthetic, sociological, and ontological clothing. I am not tempted at all to identify works with tunes or songs in the hope that I would eventually come across a completely neutral and wide-ranging concept. But the most significant difference between any other approach and my own is that I devote considerable time to sorting out the implications of describing the work-concept in terms characteristic of late eighteenth-century thought. If the central claim is correct, what conclusions are to be drawn about musical practice prior to this time? What kinds of understanding did musicians have? How did they think of their musical production? One way to reach some answers is to identify conditions of musical production that were not just alternative to those later associated with work-production, but which helped to exclude or forestall the functioning of the work-concept, not deliberately of course, but from a perspective given by hindsight. Thus the first task is to find out whether there were regulative concepts that once overruled the regulative function of the work-concept, assuming (which we should not) that the work-concept would otherwise have been functioning during this time.

5

Musical Meaning:
From Antiquity to the Enlightenment

> [I]f all the arts hold their esteem and value according to their effects, account this goodly science not among the number of those which Lucian placeth without the gates of hell as vain and unprofitable, but of such which are *pegai ton kalon*, the fountains of our lives' good and happiness. Since it is a principal means of glorifying our merciful Creator, it heightens our devotion, it gives delight and ease to our travails, it expelleth sadness and heaviness of spirit, preserveth people in concord and amity, allayeth fierceness and anger, and lastly, is the best physic for many melancholy diseases.
>
> (Henry Peacham, 1622[1])

It is not implausible to view the history of Western music as a struggle on the part of musicians to have their practice regarded as a bona fide part of whatever at a given time counted as good, serious, or civilized living. From the earliest times, theorists have distinguished respectable and valuable activities from ones judged merely entertaining and pleasurable. Usually the distinction has forced a division between what we now label élite and popular values. John Calvin was just one of many to apply the distinction directly to music. He decreed in his *Geneva Psalter* of 1543 that songs must have 'weight and majesty' and not be 'light and frivolous', for there is 'a great difference between the music one makes to entertain men at table and in their homes, and the psalms which are sung in the Church in the presence of God'.[2] Whatever material changes in musical and political hegemonies have taken place, in whatever terms the basic distinction between respectable and unrespectable music has been expressed, the distinction has never for a moment been discarded.

---

[1] From *The Complete Gentleman: The Truth of our Times and the Art of Living in London*, ed. V. B. Heltzel (Ithaca, 1962), 116.

[2] O. Strunk, *Source Readings in Music History: From Classical Antiquity through the Romantic Era* (New York, 1950), 346.

Looking at the history of music in this way prompts a strategy for describing the emergence of the work-concept; that is, we can undertake that task by tracing the development of the concept of serious music. Such a strategy is appropriate because, since 1800, serious music has come to be known as classical music and to be packaged in terms of works. The easy connection thus provides a prescription: to be a serious musician one creates or performs works. The complex connection rests upon subtle manœuvres that have taken place within theoretical and terminological debates. The result is that serious, classical, and 'work' music have come to be equated and, together, have come to be accorded the highest status possible if not in Western then at least in European musical culture.[3]

Consequently, the evidence I shall now present for the central historical claim concerns the formation of what nowadays is judged, with or without justification, to be our serious music. To find that evidence we need to look at the development of various aesthetic concepts and their surrounding contexts of theory. We also need to look at the move from extra-musical to purely musical criteria of value and classification, at the emancipation of musical sound from poetic and religious texts, at the rise of absolute or purely instrumental music, and, finally, at the articulation of the concepts of fine art and the autonomous work of art, and the subsequent inclusion of music under these categories.

At its culminating stage, the entire description is based on idealist, formalist, and romantic theories of art produced at the end of the eighteenth century. For convenience, I shall refer to this whole body of theory as *romantic*. From now on, therefore, I shall speak of my central claim in terms of the emergence within romanticism of the regulative concept of a musical work. Naturally I recognize that romanticism is an extremely complex body of theory and that my treatment of developments in aesthetic theory will inevitably be selective and partial. But I am interested only in describing those aspects of aesthetic theory (and later on, practice) relevant to the emergence of the work-concept, and certainly not every nuance and tenet of romanticism was relevant.

---

[3] From this perspective, one begins to understand why the philosophy of music and the history thereof has mostly concerned itself with 'serious' music. It is because philosophers and musicians through the ages have spent so much time concerned with the establishment of a concept and practice of music that merited the kind of attention accorded to all 'serious' or 'civilized' pursuits.

I

Of all the changes in meaning the concept of serious music has undergone throughout its history, none has been more far-reaching in its effect than that which moved musical understanding away from 'extra-musical' towards 'musical' concerns. Before 1800 the pivotal question in philosophical thought about music, 'what is music?', asked for specification of music's *extra-musical* function and significance. Music was predominantly understood as regulated by, and thus defined according to, what we would now think of as extra-musical ideals. Then, of course, they were regarded as constitutive of the musical. Such ideals affected everything musical—the theory, the conditions of production, the forms of criticism and appreciation. Usually they were shaped by the functions music served in powerful institutions like the church and court.

Those who sought to describe the nature of music looked mostly at music's ritualistic and pedagogical value. How could music successfully acquire an acceptable moral, political, or religious status that would render its production a valuable contribution to a good life? This question proved extremely difficult to answer. Practising musicians permanently found themselves in an awkward position, just because theorists were continually at loggerheads with one another as they attempted to justify the production of particular kinds of music according to extra-musical criteria. What was the problem theorists faced? Why was it so difficult to articulate the function and significance of music in extra-musical terms? To answer these questions we need to describe an alternative understanding of music, since this will provide the necessary perspective for the more detailed account to follow.

Approximately 200 years ago the situation of music and musicians changed. Though the extra-musical understanding of music continued after 1800 to influence both the theory and practice of music, its influence was marginalized by and incorporated into a new under-standing. With the rise of modern aesthetics and the fine arts, music as an art took on an autonomous, musical, and 'civilized' meaning; it came to be understood on its own terms. The basic question 'what is music?' was treated in connection less with extra-musical ideals than with what came to be regarded as 'specifically musical' ideals.

The shift from the extra-musical to the musical was marked by another shift. The question about the nature of music came to be treated through the more specific question, 'what is a musical work?' The latter question found its significance as a result of just that curious transformation which turned music into a predominantly musical affair. The transformation gave rise to a new view of music as an independent practice whose serious concerns were now claimed to be purely musical. The emerging practice became specifically geared towards, and evaluated in terms of, the production of enduring musical products. It was only with the rise of this independent conception of music, in other words, that musicians began to think predominantly of music in terms of works.

The shift from the extra-musical to the musical was marked, then, by the emergence of the concept of a musical work. But what did the emergence involve? The short and crude answer is everything that concerns musical practice—aesthetic theory about music; the music produced; the social status of musicians, be they composers, performers, or listeners; the rules, manners, codes, and mores. The long answer begins with a description of what to most readers should be some very familiar shifts in the meaning and function of music that have taken place during the course of its long history. Note, however, that though such facts might be very familiar, the conclusion drawn from them is probably and hopefully less so.

II

Sextus Empiricus, writing in the third century of the common era, captures part of an early and predominant concept of music by informing us that 'music' connotes 'a *science* dealing with melodies and notes and rhythm-making' and a *skill* 'as when we describe those who use flutes and harps as musicians'. The term is also sometimes used to refer to the *correctness* of some performance or labour. 'We speak', he reminds us, of 'a piece of painting, and of the painter, who has achieved therein correctness, as "musical"'.[4]

This description captures much of the breadth of an early concept of music, but not its full breadth. The concept expressed, for example, by the Greek term '*mousikē*' embraced any skilled activity inspired by the Muses, activities ranging from different forms of singing and dancing to those of story-telling, lyrical or poetical recitation, and theatrical performance. One of the most common activities looked

---

[4] Tr. R. G. Bury, quoted in W. Tatarkiewicz, *History of Aesthetics*, ed. J. Harrell (Warsaw, 1970), i. 227.

rather like opera. It involved three elements—word, sound, and physical gesture. It was a poem–song–dance or a physically and verbally expressive *Sprechgesang*, a heightened, dramatized speech of indeterminate pitch. It was performed by actors, musicians, dancers, or a chorus, and was put together, though not usually written down, by persons more often called poets than musicians.[5]

Musical activities of Antiquity were expressive performances rather than productive or mechanical ones. Their expressive potential was directed towards producing not a physical construction or product but the activity or performance itself. Their purpose was skilled movement, skilled instrumental playing, or expressive vocal utterance. Yet these were not their ultimate or even their penultimate purposes. The imitation or transformation of reality (*mimesis*) and the purification of souls through emotional effect (*catharsis*) were their penultimate purposes, their final purpose being the nobility of character and conduct (*kaloskagathos*) of all persons involved.[6]

Musical activities generally took distinct forms to correspond to various social activities. There were performances of tragedies as part of moral instruction. At other times, tales of noble deeds were recounted to encourage politically and socially acceptable action. Musical performances sometimes displayed a less serious content when involved in competition (*agon*) and other public and private forms of entertainment and leisure. Whatever form these activities took, their degree of acceptability and value was judged according to principles, having less to do with thoughts of imagination and creativity than with rational notions of conformity to public rule, canon, custom, or tradition. If any role was accorded to the former notions, either it was described with regard to a power of 'inspiration' granted to a poet by a higher authority to be used for appropriate ends, or it

---

[5] According to H. M. Schueller, this activity recalled a Greek legend about Harmonia according to which music, poetry, and dance exist in an aesthetic state of harmony which is the first principle of universal law (*The Idea of Music: An Introduction to Musical Aesthetics in Antiquity and the Middle Ages* (Kalamazoo, Mich., 1988), 1). The term 'chorus' designated either a group of singers or dancers, their performance, or the place of their performance, and *orchesis* was used to refer to the dance but not to groups of musicians who sang songs or played instruments (See A. Barker (ed.), *Greek Musical Writings, i. The Musician and his Art* (Cambridge, 1984), 23.) There were two terms for a poet: *aoidos* (singer) and *poietes* (maker). According to J. A. Winn, use of the former term predates that of the latter (*Unsuspected Eloquence: A History of the Relations between Poetry and Music* (New Haven, Conn., 1981), 3).

[6] Mimesis and catharsis were not always treated as distinct, but distinguishing them here helps stress different aspects of the total effect of *mousikē*.

was viewed in the context of a Dionysian or Bacchanalian performance which, unlike its opposite—Apollonian performance—represented the magical, the irrational, and the uncontrolled.[7]

On the increasingly more dominant, Apollonian side, emphasis on conformity was rigid and imperious. Theorists thought more about mimetic principles, alongside didactic ones, than any other aspect of musical performance. Evaluation focused on correctness of execution— how well appropriate extra-musical rules or principles were followed. Such rules were formulated in terms regarding the imitation of actions, events, and characters, and the moral effects of a performance. The amount of pleasure experienced, on the other hand, was rejected as a suitable sole criterion of worth. Novelty, virtuosity, and almost any form of personal expression or gratification were deemed inappropriate unless carefully tailored to bring about beneficial ends.[8]

Thus Plato informs us that 'those who are looking for the best kind of singing and music must look not for the kind that is pleasant but that which is correct: and . . . an imitation is correct if it is made like the object imitated, both in quantity and quality'. Aristotle also makes it quite clear that the poet should not speak '*in propria persona*', because he does not imitate if he does that. And when it was asked why people listen with more pleasure to songs they happen to know already, this answer was readily available: 'It is because it is more obvious when the singer . . . hits the target.' Plutarch was later to make the point against novelty and personal creativity in the sharpest of terms:

Crexus, Timotheus, and Philoxenus . . . and other poets of the same period displayed more vulgarity and a passion for novelty, and pursued the style nowadays called 'popular' or 'profiteering.' The [unfortunate] result was that music, . . . simple and dignified in character, went quite out of fashion.[9]

---

[7] See Plato, *Laws*, tr. T. J. Saunders (Harmondsworth, Middx., 1970), II: 652–701, pp. 85 ff.

[8] See Plato, ibid., and Aristotle's discussion of the use and potential misuse of pleasure and play in the education of children in his *Politics*, VIII: 4–7 (McKeon (ed.), *The Basic Works of Aristotle*, 1305 ff.). Note that conformity to rules is not antithetical to the improvisatory nature of performance. Of necessity, performers used a combination of memory (repetition) and novel improvisation to continue and extend what was essentially an oral tradition.

[9] Barker (ed.), *Greek Musical Writings*, 152; Aristotle's *Poetics*, 1460a: 7–9. The Aristotelian *Problems* (Barker (ed.), *Greek Musical Writings*, 190); Plutarch (Barker (ed.), *Greek Musical Writings*, 218). Plutarch is not speaking here of what I earlier referred to as operatic activity, but of a simpler form of music. The same points apply, however.

Preference for strict conformity stemmed from the belief that muse-inspired activities were inextricably associated with education (*paideia*); their significance extended far into the moral and political domain. Proper musical performance was never thought to be appropriate first and foremost for entertainment or leisure, because it was capable of representing society's values at large. In a mostly illiterate society, songs and melodies could be used—as Terpander and, later, Plutarch demonstrated—to embody particular statements of law, custom, or convention, ranging from the political to the scientific. Plato expressed the prevalent thought most directly. Persons should fear, he wrote, when anyone says that

Men care most for the song, which is newest to the singer, lest anyone should praise this, thinking that the poet meant not new songs but new ways of singing. One should not praise such a thing nor take it up; one should be cautious in adopting a new kind of poetry and music, for this endangers the whole system. The ways of poetry and music are not changed anywhere without change in the most important laws of the city.[10]

Not only change of musical form or style, but personal expression as well, could threaten whatever social and political stability a city possessed, even or especially Plato's delicately balanced city of guardians, soldiers, and ordinary citizens.

Conservative musical activity had a positive value precisely, then, in its connection to the idea of a good or educated life. Plato wrote:

the man who can best blend gymnastics with music and administer them, perfectly measured, to the soul, is the one whom we should most correctly call the complete musician and the true expert in harmony, much more than the man who can tune strings to one another.

That the guardians of his ideal city had 'harmonious' or 'well-attuned' characters was a belief entirely supportive of Plato's general doctrine of *ethos*, a doctrine encouraging a kind of moral or political indoctrination, in part by using the medium of music.[11]

---

[10] *Republic*, IV: 424c (tr. G. M. A. Grube (Indianapolis, 1974), 90). Cf. W. J. Ong, who writes that 'The Muses . . . not only typify the role of sound in an oral culture but also advertise the place of celebration and pleasure in the pursuit of knowledge such as oral cultures possess: they belong to the era before the pursuit of knowledge took on the trappings of "work," [labour] and indeed to an era when work and play are clearly less distinct that they are in an alphabetic, technologized culture' (*The Presence of the Word: Some Prolegomena for Culture and Religious History* (Minneapolis, 1967), 70).

[11] *Republic*, III: 412a (Barker (ed.), *Greek Musical Writings*, 128–9). Recall, also, Plato's scepticism regarding musical activity. Concerned, as it was, in its mimetic role

Plato's pronouncements reveal that he thought of music in the first place a matter of theoretical speculation. A musician was at best a philosopher or scientist and at worst (merely) a practitioner. Centuries later, Boethius was inspired to draw the same conclusion. Distinguishing in Platonic terms three levels of music: cosmic harmony, harmony of human body and soul, and harmony of song and melody, he claimed that the first alone is the proper subject of our speculation; it is the one with complete value to which the others can only ever approximate. 'The learning of music in rational understanding' is much more perfect 'than in production', Boethius thus argued.

A musician (*musicus*) is a man who, by his reason, has engaged in the science of music, not in order to practice it, but from speculative interest. . . . A musician . . . possesses an ability, based on rational theory and in accordance with music, to judge melodies and rhythms, as well as all kinds of song and the verses of poets.[12]

This view was prevalent in even later centuries as well. Drawing an analogy with typically derogatory connotation, John Cotton of the twelfth century concluded that 'the singer (*cantus*) is like a drunkard who does indeed get home but does not in the least know by what path he returns.' And in the eighteenth century, Roger North reminded his readers that, though the 'execution and invention' of music might be an art, composition is a science which 'all learners' should develop their 'judgement to know'.[13]

Such a view seems to have been taken wholly for granted in Antiquity. The relevance of moral notions in determining the significance of music rarely found itself expressed in reference to musical production, but through all manner of theoretical or scientific speculation. Consider the physiological/psychological account expressed in the Aristotelian *Problems*. There, it is argued that the movement inherent in melody—in the ordering of high and low notes, though not in concords—is 'related to human actions', and that such movement conveys moral character.[14] The relation suggested here between musical sounds and moral character, established through our hearing sounds-in-their-relation or in-their-movement, closely matched the

with pretence, opinion, and semblance, of what use could it be, he asked, in our search for genuine knowledge?

[12] See Strunk, *Source Readings*, 84–6, and Tatarkiewicz, *History of Aesthetics*, ii. 87.

[13] See Schueller, *The Idea of Music*, 364; *Roger North on Music: Being a Selection from his Essays Written during the Years* c.1695–1728 ed. J. Wilson (London, 1959), 24.

[14] *Problems*, xix. 27 (Barker (ed.), *Greek Musical Writings*, 197).

influential Pythagorean thesis concerning the special nature of musical intervals. Pythagoreans had argued that consonant, intervallic structures of music, analysed in terms of the ratios between the lengths of stretched strings and their vibrations at certain pitches, mirrored corresponding ratios found in the moral character (the human soul) as well as in the astronomical system of heavenly bodies. The beauty, significance, and power of music as an art of sound found its source and explanation precisely in this mirroring relation.

Speculations such as these provided theorists with two perspectives from which to assess the music actually produced. One could either look to the patterning inherent in music *qua* sounds-in-their-relation to find indications of moral character, or one could assess the moral effects of a musical production in which a melody accompanied a text. Though a choice existed, this description underemphasizes a crucial imbalance. For the most part it was believed that the words of a text captured music's meaning much more adequately than sounds by themselves; sound and dance movements were usually relegated to the status of accompaniment either to a text or to an 'occasion' which would provide them, by association, with meaning. The use of words, however, was often considered essential to any musical occasion if that occasion was to be regarded as edifying, truthful, and thereby respectable. The immediate implication of this belief was that music without words—what we now call (purely) instrumental music—ended up being rejected on the grounds that, by itself, it had no, or at least insufficient, moral import and, therefore, was probably of very little import at all.

It was quite consistent, then, for theorists to reject the perspective of the post-Epicurean philosopher Philodemus, who argued that music could be judged according not to the moral value of associated words, but, rather, to the pleasure it promoted. He maintained that musical sounds when allied to poetry tended to mislead theorists into thinking that the sounds themselves have moral or social meaning. To assess the real significance of music, he suggested, one should look at sounds-in-their-relations *independently* of the word. For then one would find that music is in fact a source of harmless pleasure, the significance of which points to nothing beyond the sounds themselves or the pleasure experienced.

No melody, qua melody, being irrational (*alogon*) . . . rouses the soul from a state of tranquillity and repose and leads it to the condition which belongs naturally to its character. . . . For music is not an imitatory art (*mimetikon*) . . . nor does it . . . have similarities to moral feelings . . . any more than cookery.[15]

The basic drift of Philodemus' argument would be accepted only centuries later. At the time his argument (or any comparable view) would be mistakenly interpreted as further supporting the belief that melodies without words lacked sufficient moral or social meaning, and thereby any clear positive value, for their production to be further encouraged.

That assessment did not result in instrumental music's ceasing to be produced. It meant only that when it was produced, it was given a subsidiary role in public performance. Such music found its place, occupying prelude or interlude space, before or during a vocal or text-based performance. It was further used as a background accompaniment to 'give the beat' on social, political, or military occasions. It was used abundantly for private purposes of practice—for the exercising of one's fingers and the general development of technique. Otherwise it existed as a sort of popular or folk music.

Given these different sorts of uses, instrumental music tended not to be treated as a unified body of production or practice. Instruments themselves were usually individuated according either to the kind of music they produced (functionally defined) or to the moral qualities associated with the players. The manner and content of one's musical performance were considered as much an indication of one's moral character (or lack thereof) as those of any other human activity. Tatarkiewicz once concluded that 'the Greeks regarded the two types of instrument [the aulos and the lyre] as so different that they did not even include all instrumental music within one concept'.[16] He overstated the point.

Instruments were generally categorized under concepts more specific, or quite other, than the generic concept of instrumental music. This does not mean, though, that theorists did not speak about instrumental music as a particular kind of music in any terms at all. Evidence suggests only that what we now call an instrument's musical significance was usually, but not always, overridden by or subordinated to an instrument's extra-musical significance. Plato famously rejected the aulos from his ideal city because it encouraged complex music and

---

[15] Quoted in M. C. Beardsley, *Aesthetics from Classical Greece to the Present: A Short History* (University of Alabama, 1966), 73.

[16] *History of Aesthetics*, i. 19.

such music was considered dangerous. Aristotle rejected flute-playing because a player's face became distorted whilst playing the flute, and such distortion was unnatural.[17] But still there were mentions of instrumental music as a whole having a specific kind of (high or low) value.

The same argument would apply were it claimed that there was no unified, musical conception of modes or scales in Antiquity. Again one finds fewer references to a mode's particular musical qualities than to its distinct moral attributes and conventional emotional effects deriving from its association with a people or culture after which it was named. Even those interested in the formal classification of the modes according to intervallic difference realized this was to display only a partial interest. Extra-musical considerations were necessary as well.[18] So, for the most part but not always, the Dorian, Ionian, Lydian, and Phrygian modes were differentiated and used with thoughts of qualities, such as courage, fortitude, and, negatively, effeminacy, in mind.

All in all, what we now take to be musical matters, having to do with the nature of compositional form and performance technique, modes, melodies, and instrumental colouring, were rarely accorded importance unless they were deemed relevant to music's function as a politically powerful form of human expression. Music's natural power to affect persons was its primary attribute, and its power to affect persons beneficially its primary function. Consequently, music had first and foremost to be scrutinized in socio-political terms, and, if and when appropriate, carefully censored.[19]

For our purposes, the most significant consequence of placing value on the moral and educational role of music was that music found its sister disciplines, not in what nowadays we call the plastic arts—notably painting and sculpture—but in the liberal arts of rhetoric, dialectics, and grammar. And then, under the additional influence of Pythagorean theory and the like, music was classified alongside the arts of arithmetic, astronomy, and geometry. Such classifications helped to obscure (from a modern perspective) connections that would later be made between music and the constructive, mechanical, or plastic arts. This fact has essential bearing on the development—or

[17] Plato, *Republic*, III: 399d; Aristotle, *Politics*, VIII: 1341a 17–b19.

[18] I am adapting a thought expressed in the Plutarchian Treatise, that once a musician has learned the elements of musical composition, he must learn how to distinguish the morally 'appropriate' from the 'inappropriate' (Barker (ed.), *Greek Musical Writings*, 241).

[19] Cf. Plato, *Laws*, II: 669a.

lack of it—of a work-concept in the field of music, a matter to which I shall return independently in the next chapter.

This basic classification of music extended well past Antiquity. In the thirteenth century, Dante described musical-poetry as 'a rhetorical composition set to music'. And well into the eighteenth century the German flautist Johann Joachim Quantz reiterated the musician's responsibility to guide and influence the moral character of society.[20] For at least this long, music remained a liberal art. In fact, as we shall now go on to see, remarkably little changed in this basic, 'extra-musical' understanding of music between Antiquity and the Enlightenment.

III

With the rise of Christianity, theoretical speculation about music altered insofar as 'serious' music now had to serve God before all else. This alteration was marked by the introduction of new forms of musical presentation. Sacred or plain chants and hymns, as distinguished, say, from the songs, dances, and performances of more local, more secular, and more 'mythic' rituals and festivals, were introduced into the church as integral parts of the service. Constitutive of the new 'serious' music, these forms supplanted the element of theatre characteristic of ancient musical performance; they were stripped of the element of dance and, later, sometimes even of instrumental playing. They were sung either by a congregation or chorus, or by both. Often hymns were accompanied by hand-clapping, but that eventually ceased as well. Most church music was sung initially in unison, later in parts, but—and this was their most significant feature—in a declamatory style designed to give priority to the word, the essential medium in religious devotion.[21]

---

[20] Dante, *De vulgari eloquentia*, tr. A. G. Ferrers Howell (London, 1890), II. 4, 'On the Variety of Style of Those who Write Poetry', 55; Quantz, *On Playing the Flute* (1752), tr. E. R. Reilly (New York, 1985), 11 ff.

[21] The presupposition behind decisions regarding the style of church music was that some kind of music was needed or necessary in the first place. Cf. Augustine, who approves the custom of singing in the church, *Confessions*, tr. R. S. Pine-Coffin (New York, 1961), x. 33: 239. Note also that 'Musica' was a term rarely used with reference to the music of the Christian church; 'psalm', 'chant', 'hymn', 'conductus', 'motet', (and foreign equivalents) were the more common terms. 'Musica' tended to be reserved for knowledge of music rather than for its practical activity or production. See Schueller, *The Idea of Music*, 364 and Winn, *Unsuspected Eloquence*, 38 and 72.

The gradual transition first from simple to more complex monophony, and then from monophony to polyphony, threatened the
required declamatory style. Complexity of melody and rhythm was
thought to distract worshippers from their prayers. Rather than
encouraging 'devotion in the heart', it was said, 'complex' melodies
'stir lascivious sensations in the loins'.[22] In the fourth century,
St Jerome expressed his concern when he asked Christians to 'sing to
God, not with the voice, but with the heart; not after the fashion of the
tragedians, in smearing the throat with a sweet drug, so that theatrical
melodies and songs are heard in the church, but in fear, in work, and
in knowledge of the Scriptures'. Though a person might be '*kako-
phonos*', he continued, 'if he have good works, he is a sweet singer
before God'. But that was not all. His further restriction for 'the
servant of Christ' was that he should not sing 'through his voice, but
through the words which he pronounces, in order that the evil spirit . . .
may depart from those . . . who would make of the house of God a
popular theatre'.[23]

Since it was believed that the words alone carried religious meaning, it
was held that complex tonal structures, by distracting Christians from
the word, necessarily undermined the religious character of music.
This belief was supported by a claim reminiscent of the Pythagoreans,
that complex structures of sound failed to harmonize with the structure
of the heavens. Few could accept, then, Johannes Scotus Erigena's
argument offered in the ninth century in favour of the increased
complexity of musical sound, be it 'natural' (cosmic) or instrumental
(artificial). He agreed with earlier theorists that the harmonies and
rhythms of humanly produced music, and the internal, mathematical
relationships of audible objects could be equated with the invisible,
musical and mathematical relations of the cosmos. He then suggested
that if cosmic order is itself complex and multi-dimensional, those
audible objects embodying the same numerical principles as the
cosmos could be complex and multi-dimensional in comparable ways.[24]

Increase in harmonic and contrapuntal complexity within the
liturgical chant seemed unavoidable, and it was just this movement
towards giving musical sounds a greater role to play in musical
practice that motivated theorists—Guido d'Arezzo in the eleventh

---

[22] John of Salisbury, quoted in Schueller, *The Idea of Music*, 356.
[23] Strunk, *Source Readings*, 72.
[24] For versions of this argument, see Sparshott, 'Aesthetics of Music', *The New
Grove Dictionary*, i (London, 1981), 124 and Schueller, *The Idea of Music*, 294–5.

century and Franco of Cologne in the thirteenth—to seek systematic musical notations. A philosophical justification for the use of notation (dubious as it was) was forthcoming—though when used to justify the production of complex music and notation, it found limited acceptance. By mirroring the pattern of music, it was claimed, notation indirectly symbolized the harmonious structure of the heavens.[25]

Reflecting the predominant belief that tonal structures were meaningful only when mediated by a text, usually a religious one, most notations incorporated literary or poetical signs in addition to tonal and rhythmic ones. Melodic patterns were usually designed to follow the 'natural' rhythm of the words, so that the length of individual notes would correspond precisely with the time it took to utter each word's syllables. This rendered duration signs for musical notes unnecessary. Though the introduction of complex and later polyphonic structures complicated the relation between melody and text, necessitating in turn new notational devices, many 'musical' signs were introduced into notation and then standardized only when music liberated itself completely from the text. But that was many centuries after the Middle Ages.[26]

In line with the emphasis on the word, theorists believed for a long time that the human voice was the only *pure* musical instrument, the only instrument appropriate for articulating the Word of God. The voice alone conveyed rational or natural music and thus was regarded as an appropriate route to God. Most 'serious' music, therefore, was exclusively vocal or choral. Despite post-Pythagorean attempts to find evidence for the significance of instrumental music, this music continued to be regarded by many as 'impure', 'empty', and 'bare' and to be relegated to the practice of the 'rustic and uneducated', to persons with 'empty minds'.

'Types and figures aside, whence . . . all these organs, all these cymbals, in the church?', had been Aelred of Rievaulx's question in the twelfth century, as he even sought to banish organ-playing from the church. 'To what purpose', he had asked, 'that horrible inflating of bellows, expressing the crashing of thunder rather than the smoothness of human voice?'[27] Aelred's contempt was no worse than Boethius's had been. Boethius had set the standard of contempt for the music

---

[25] Cf. Sparshott, 'Aesthetics', 124.
[26] Cf. Schueller, *The Idea of Music*, 317, and G. Abraham, *The Concise Oxford History of Music* (Oxford, 1979), 85.
[27] Schueller, *The Idea of Music*, 354.

of instruments centuries earlier, when he declared with cutting nuance that 'one kind of music uses instruments and another produces poetry'.[28]

Because it was widely accepted, moreover, even in the later Middle Ages, that a text could and should influence the composition of an accompanying musical structure, it was rarely acknowledged that the influence should or even could work the other way round. When it was, as it increasingly came to be, the acknowledgement was put in the form of a complaint, that words are no longer treated as primary but as mere pretexts for musical composition (narrowly conceived as a composition of sounds). In practice, music was being set to pre-written texts; sometimes, however, pre-written texts were being set to pre-composed melodies, melodies that might either have been composed previously for another text or as part of an instrumental composition. Even though adaptation to a text normally required substantial changes in any 'pre-existing' melodic structure, the very idea of a text's being set to music was sufficient to worry the authorities. The word had traditionally come first, and it should continue to do so— not only in theory but in practice as well.

That musicians of this period were rarely confined to the composition and performance of a single type of music helps explain another feature of their practice, namely, that there was a significant overlap in style and content between different types of music. Although church music was considered by far the most acceptable form of music, having apparently been composed in accordance with the highest religious principles, developments on the outside—in secular music—managed to influence the composition of much sacred music. In other words, despite the strict segregation desired between sacred and secular music, the latter often affected and shaped the former, certainly in matters of musical form if not in literary content. Developments in secular instrumental music directly affected developments, for example, in the polyphonic forms of church music, because secular compositions (or parts thereof) were transformed into sacred ones through borrowing, recomposing, or incorporation. Again, such transformational procedures were rarely publicly acknowledged by the authorities.

As it turned out, however, incorporation of instrumental music into a sacred performance proved a particularly successful way for musicians

[28] Tatarkiewicz, *History of Aesthetics*, ii., 87. Cf. Dante's description, that no 'trumpeter or organist or lutenist calls his melody a song, except insofar as it is married to some [cantioni]' (Winn, *Unsuspected Eloquence*, 85).

to produce instrumental music of an acceptable sort. Instrumental music could be performed on its own, the authorities said, but then only in secular and, even better, in private quarters. In public it was to be heard in the context of, and as a part of, a text-based performance constitutive of a religious service, say, or a royal celebration. For so contextualized, such music could acquire or derive a respectable meaning by virtue of its being part of an acceptable whole.

Musical practice of the Middle Ages continued, finally, to be basically conservative for reasons comparable to those given in Antiquity. Music demonstrating newness, personal innovation, or creativity, for example, was valued only if it strictly conformed to the traditions of the church—the teaching body. An impossible demand? At times it seemed so. Practising musicians had to reconcile their taste for variety and innovation with their social and religious obligations. The more talented survived somehow, but rarely without interference from above. In 1324 Pope John XXII issued a bill forbidding the use of new music in the Mass. 'Originality', he had decided, 'encourages effeminacy in descant, a rushing on without rest, and an intoxication of the ear without the healing of the soul.'[29]

But as I have already mentioned, creative developments in composition took place none the less—despite the church's demands. Affecting the composition of both instrumental and vocal music, new stability was given, in the fourteenth-century compositions of the *Ars nova*, to principles of musical closure (of open and closed endings of melodic formulae), repetition, and ornamentation. Such principles were increasingly conceived independently of corresponding textual or poetic principles, as well as in isolation from specific religious and other extra-musical demands. But they were not as yet conceived in complete independence.[30]

IV

The Renaissance witnessed an explicit revival of classical views which disputed the need for music predominately to serve religious ends. Music could and should serve secular ends as well. The resulting humanistic conception of music would probably have been much more effective in altering attitudes towards music had the Renaissance not been followed by the Reformation.

[29] Schueller, *The Idea of Music*, 412.
[30] Abraham, *The Concise Oxford History of Music*, 70, 72, and 119 ff.

The emphasis on secularity was articulated by philosophers who thought that music's most important role is as a liberal art. If music has any role in human affairs, it is as a language capable of conveying, mirroring, or affecting human and social qualities. This mirroring is not a mediating function between persons and their object of worship, but is directly functional in forming a person's understanding of his or her moral life. Vincenzo Galilei, in 1581, expressed this view most directly when he wrote that 'if a musician has not the power to direct the minds of his listeners to their benefit, his science and knowledge are to be reputed null and vain, since the art of music was instituted and numbered among the liberal arts for no other purpose'. Such a view was further supported by the belief that the ever more complex structures of musical counterpoint and harmony are human, rational constructs, albeit designed according to natural and mathematical principles.[31]

This view did little to improve the status of instrumental music, however. Despite major contributions to theories of composition—to harmony and counterpoint—humanists continued to think that music conveys human passion, or influences a person's moral character, essentially by means of the word. How, without words, could music direct the minds or characters of the listeners? Melodies alone might give rise to sensory pleasures and they might trigger our emotions, but how could such pleasure or feeling—as Plato and Aristotle too had asked—be of moral or educational significance unless it was controlled externally by the word and thereby subordinated to some other valuable effect? The assumption of the time was clear. Melody lacking reference to anything valuable beyond itself had no acceptable function.

Even Gioseffo Zarlino, renowned for his contribution in the sixteenth century to developments in instrumental music, was convinced that 'harmony and rhythm ought to follow speech'.

For if in speech, whether by way of narrative or of imitation (and these occur in speech), matters may be treated that are joyful or mournful . . . we must also make a choice of a harmony and a rhythm similar to the nature of the matter contained in the speech in order that from the combination of these things, put together with proportion, may result a melody suited to the purpose.[32]

---

[31] Strunk, *Source Readings*, 319 and Sparshott, 'Aesthetics', 124.
[32] Strunk, *Source Readings*, 255–6.

As he acknowledged himself, he was adopting Plato's view, lock, stock, and barrel.

As a new response to an old view of music, several musicians— among them, Galilei—joined together to form what came to be called the Florentine Camerata. Openly influenced by Greek tragedy, they sought to give the latter contemporary expression. The result was a dramatic musical recital, known as the *stile recitativo*. Those who supported this sort of music respectfully dismissed alternative styles in choral or instrumental composition, especially music that contained over-complex, contrapuntal structures. Galilei pointed out:

> For all the height of excellence of the practical music of the moderns, there is not . . . the slightest sign of its accomplishing what ancient music accomplished, nor do we read that it accomplished it fifty or a hundred years ago when it was not so common and familiar to men. Thus neither its novelty nor its excellence has ever had the power . . . of producing any of the virtuous, infinitely beneficial and comforting effects that ancient music produced.[33]

The prevalence of conservative attitudes goes a long way towards explaining the poor reception given to Palestrina's *Missa Papae Marcelli*. When first performed the music offended Pope Gregory XIII, who failed, it is said, to grasp the 'absurd and unmeaning complication of the sounds'. The Pope threatened to banish the mass for the same reasons he banished any music 'overflowing with barbarisms, obscurities, contrarieties, and superfluities . . . [resulting from] the clumsiness or negligence or even wickedness of the composers, scribes, and printers'. Palestrina, though required to follow the decrees of his employer, appealed. On the second hearing, the Pope accepted the mass for its 'grave and dignified simplicity'.[34]

---

[33] Strunk, *Source Readings*, 290–302 and 306.

[34] See Hogarth, *Music History*, 24 and Strunk, *Source Readings*, 357. Agostino Agazzari provided a contemporary comment. Music of an ancient kind, he wrote, 'is no longer in use, both because of the confusion and babel of the words . . . and because it has no grace, for, with all the voices singing, one hears neither period nor sense. . . . And on this account music would have come very near to being banished from Holy Church . . . had not Giovanni Palestrina found the remedy, showing that the fault and error lay, not with music, but with composers, and composing in confirmation of this the mass entitled Missa Papae Marcelli. For this reason, although such compositions are good according to the rules of counterpoint, they are at the same time faulty according to the rules of music that is true and good, something which arises from disregarding the aim and function and good precepts of the latter; such composers wishing to stand solely on the observance of canonic treatments and imitation of the notes, not on the passion and expression of the words' (Strunk, *Source Readings*, 430–1).

Not only in the Catholic church but in the newly reformed Protestant churches as well, the traditional religious demand for declamatory style retained its regulative force. No one better demonstrated this fact than Calvin. 'When we sing [the psalms of David],' he assured his readers, 'we may be certain that God puts the words in our mouths as if Himself sang in us to exalt His glory.' The melody of the song, therefore, should 'be moderated', and be of a certain design, so that it has a 'weight and majesty proper to the subject'. Then it may even be suitable, Calvin concluded, for 'singing in church'.[35]

Most writers and thinkers, whether humanist or ecclesiastic in their orientation, continued to subordinate the composition of melody to the text. No debate in this time, however strenuous—and the most strenuous was undoubtedly between Galilei and Zarlino—was able to secure for purely instrumental music any acceptable value. Thus, in 1600, instrumental music of the modern sort was concluded to be 'absurd!'[36]

Yet if this attitude reigned in public, there were also undercurrents or 'avant-garde' forces flowing beneath the surface. Some thinkers, who wished to explore the possibilities of instrumental music, even if they ended up actually producing music with words, appealed to that originally Epicurean idea (expressed by Philodemus) that the beauty of music should be judged not according to its mimetic or referential qualities, but rather with respect to the pleasurable and passionate responses triggered solely by auditory sensation. Zarlino and Monteverdi, for example, considered this view. Considerations like theirs helped form an alliance between instrumental music and a theory of emotional expressiveness in music that, much later, would be judged to have contributed significantly to the formation of an independent and purely musical practice. However, that development could take place only when, among other things, the principles of pleasure and expression, reinterpreted to stand as marks of aesthetic intuition and taste, gained recognition as appropriate criteria of musical worth. But in hindsight the fundamental hint had already found its place in musical history: instrumental music needed to find a meaning and worth that was not already applicable to, or straightforwardly derived from, another expressive medium.

During the Renaissance, however, music retained its status as a liberal art produced to edify all persons involved. Overlapping with

[35] Strunk, *Source Readings*, 348.
[36] G. M. Artusi, in Strunk, *Source Readings*, 396.

that status, it remained a medium for religious devotion within the ever-powerful church. The composition of music remained, furthermore, regulated by the function of words.

V

The same conclusions can be drawn as we approach the seventeenth and eighteenth centuries. Lasting until the end of the eighteenth century in fact, the view of music present in Antiquity and then in the Middle Ages continued to exert its power to the unhappiness only of musicians of 'modernist' inclination. Musical understanding continued, therefore, to refer to activities guided by and subordinate to the needs of the educational authorities, at this time still the church, and, usually in close association with the church, the scientific establishment.

Strict religious ideals were shown, for example, in J. S. Bach's terms of employment in Halle. His musical activity, he was instructed, should always 'be conducted in such a way that the members of the congregation shall be the more inspired and refreshed in worship and in their love of harkening to the Word of God'.[37] Influential philosophical, scientific, or theoretical speculation about music managed, furthermore, to maintain for the concept of music a wide-ranging application. A philosopher–scientist could still with sense be called a musician. Thus in 1618, Descartes produced a rationalistic and mathematically based conception of musical principles of acoustics and harmony in his *Musicae Compendium*, which directly influenced Jean Philippe Rameau's *Traite de l'harmonie réduite a ses principes naturels*, written a century later. In the latter, the author opens with the following observation:

However much progress music made until our time, it appears that the more sensitive the ear has become to the marvelous effects of this art, the less inquisitive the mind has been about its true principles. One might say that reason has lost its rights, while experience has acquired a certain authority.

He continues:

If modern musicians (i.e., since Zarlino) had attempted to justify their practices, as did the Ancients, they would certainly have put an end to prejudices [of themselves and others]. . . . Music is a science which should have definite rules; these rules should be drawn from an evident [self-evident]

[37] David and Mendel (eds.), *The Bach Reader*, 65.

principle; and this principle cannot really be known to us without the aid of mathematics.[38]

More influential in the actual practice of composition was J. J. Fux's *Gradus ad Parnassum* written in 1725. His system of counter-point—with specifications of the rules of musical composition still in use today—was derived, he also claimed, from rational principles. To that claim he added, however, that such rules were derived from the traditional conception of church modes rather than from the recent development of 'modern' scales. Fux accordingly instructed musicians to 'proceed to [their] work [the classification of intervals, consonances, etc.], taking [their] beginning from God himself, thrice greatest, the fount of all the sciences'.[39]

In 1734, Lorenz Mizler reproposed the ancient view that music amounted to no more than a body of scientific knowledge, the practice thereof being of near-incidental value only. To support the associated claim that the study of music is an indispensable part of the education of the philosopher–scientist, he founded in 1738 the *Societät der musikalischen Wissenschaften*, and in 1739 the 'first musical periodical', the *Neu eröffnete musikalische Bibliothek*. According to J. S. Bach, the society concentrated far too much on 'dry, mathematical stuff', though that did not prevent him from becoming one of its members.[40]

Theorists continued as well to embrace the mutually contradictory understanding of music, on the one hand, as a science of harmony and rhythm, and, on the other, as a production in which melody accompanied text. But they did not embrace this double-sided view by simply adopting views of old; their modern theories committed them to this understanding. Unfortunately, such theories did not resolve any of the tension existing within it.

In 1701 composer George Muffat was forced to admit that his first collection of instrumental concertos are 'suited neither to the church (because of the ballet airs and airs of other sorts which they include) nor for dancing (because of other interwoven conceits, now slow and serious, now gay and nimble, and composed only for the express

---

[38] *Treatise on Harmony* (1722), tr. P. Gossett (New York, 1971), pp. xxxiii–xxxv.

[39] Strunk, *Source Readings*, 537. Winn points out that Fux developed a notion of imitation in his method for part-writing, the meaning of which, though elusive, enabled musicians to justify their composing more complex musical parts within their composi-tions (*Unsuspected Eloquence*, 195). For detailed commentary on the prevalent 'scientific' view of music, see A. Cohen, *Music in the French Royal Academy of Sciences: A Study in the Evolution of Musical Thought* (Princeton, 1981).

[40] David and Mendel (eds.), *The Bach Reader*, 25.

refreshment of the ear)'. But, he continued, they 'may be performed most appropriately in connection with entertainments given by great princes and lords, for reception of distinguished guests, and at state banquets, serenades, and assemblies of musical amateurs and virtuosi'.[41] Acceptance of purely instrumental music was still limited and socially confined, despite an abundance of widely read and accepted treatises on instrumental harmony and counterpoint. But the fact that it was increasingly being accepted in civilized secular quarters, especially in the courts, allowed for a significant increase in its production. That seems to have made the tension all the worse for the theorists. For them, no amount of production or rationalist-scientific argument could convince them that instrumental music was worthy. Indeed, they found more reasons to continue to judge it completely worthless. Instrumental music ended up being accorded its traditional lowly status until late in the eighteenth century.

Yet there was a movement in the eighteenth century that helped shift music away from its traditional, extra-musical understanding and which ultimately brought about a change in the status of instrumental music. Were one to look closely in fact at the development of aesthetic theory in the first decades of the eighteenth century, and then at artistic and musical forms other than instrumental music, one would see far-reaching and major changes taking place. With all their various nuances, these changes were constitutive of the development of the fine arts. The changes prompted not so much a new way, but a new expression of a way, to assess art. This new expression sought to overhaul the traditional principle of imitation (mimesis).

One crucial alteration, or at least change in emphasis, in the principle was this. While it had previously indicated the mirroring or representation of particular passions or particular objects of nature, imitation was increasingly defined as the mirroring of the essence, ideality, or general form of these phenomena. Johann Joachim Winckelmann suggested in the 1750s that in painting, the use of allegory helps give poetic form to general ideas. But, he warned, allegory should clothe and not conceal these ideas, for it is they alone that give an artwork its value.[42] Samuel Johnson famously decreed that the poet should no longer count the streaks of the tulip; Joshua Reynolds

---

[41] Strunk, *Source Readings*, 449.

[42] *Reflections on the Imitation of Greek Works in Painting and Sculpture* (1755), tr. E. Heyer and R. C. Norton (La Salle, Ill., 1987), § vii. Winckelmann confines his attention to painting and sculpture and says almost nothing about the other arts.

likewise determined that one should focus on nature as universal and not as particular. Yet general forms or essences of natural or human phenomena could only be mirrored via concrete representation, it was believed, and thus imitation of the general amounted to imitation of the *general-via-the-particular*.

There had always been numerous ways to understand the concept of imitation, but what all its uses had shared was the positing of an external relation between the art and that which is imitated. Differences had existed not only in the nature of the object imitated, but also in the mimetic relation. Thus, imitation had been understood as 'copying or learning from the ancients', as when one modelled one's work on previous masterpieces.[43] It had been understood as recounting stores and tales, often heroic deeds; or as representing physical objects, sounds, and movements through a given medium, as when a painter paints a still life of a vase. In cases of musical imitation, one finds references as well to the imitation of form, as when a musician replicates in sound the form of verbal intonation or the structure of a text. One finds the concept of imitation also meaning exemplification, as when one exemplifies, through its use, a tonal or rhythmic model, melody, or motif. However, if we want to understand why instrumental music in particular acquired its positive value so much later than most other forms of musical or artistic production, it is the 'general-via-particular' view of imitation that we will need eventually to focus upon.[44]

In hindsight, this view of imitation was the first step in the development of what was to become a romantic conception of fine art. And, as we shall soon see, instrumental music first found its nobility within romantic theorizing on the fine arts. The second step in the romantic aesthetic was to eliminate the principle of imitation altogether. Romanticism was to replace the principle of imitation with a combined principle of expression and embodiment. Until imitation was replaced,

---

[43] Winckelmann offers a paradigmatic statement of this idea: 'The only way for us to become great, lies in the imitation (*Nachahmung*) of the Greeks.' I have discussed this claim in my review of his *Reflections*, tr. E. Heyer and R. C. Norton, *Teaching Philosophy*, 12:3 (1989), 329–32.

[44] Cf. Beardsley's discussion of the French Academy's theory of ideal imitation (*Aesthetics*, 148). Detailed expositions on musical mimesis are given by John Neubauer, *The Emancipation of Music from Language: Departure from Mimesis in Eighteenth-Century Aesthetics* (New Haven, 1986) and Winn, *Unsuspected Eloquence*. For canonical discussions of mimesis and related concepts, see E. Auerbach, *Mimesis: The Representation of Reality in Western Literature*, tr. W. R. Trask (Princeton, 1973) and P. Kivy, *The Corded Shell: Reflections on Musical Expression* (Princeton, 1980).

however, theorists continued to think of music as shaped and contoured necessarily by a text.

Early to mid-eighteenth-century discussions of the mimetic principle affected all the arts as we classify them today. They helped move them—though not entirely successfully—away from their traditional 'extra-artistic' classifications into a class of their own. Any medium that could meet the new mimetic condition acquired the status of a fine art. Though the mimetic content of the 'new' principle continued to be variously understood, for the first time painting, sculpture, architecture, poetry, and music (with words) were exclusively chosen to constitute what Paul Oskar Kristeller has called the 'irreducible nucleus of the modern system of the arts'. Music was included for reasons clearly stated by Jean Baptiste Du Bos.

The basic principles that govern music are similar to those that govern poetry and painting. Like poetry and painting, music is an imitation. Music cannot be good unless it conforms to the general rules that apply to the other arts on such matters as choice of subject and exactness of representation.[45]

Conceiving music as mimetic encouraged theorists to continue to reject instrumental music as a really valuable art. They just could not comprehend how music without words could successfully imitate, in whatever way imitation was understood. Melodies simply did not succeed in pointing adequately to anything beyond themselves, even the ideality of nature as exemplified in particular form. Whereas the newly classified arts of painting and poetry captured 'clearly and distinctly' the essence of Nature, of person or world, through their particular representational media, instrumental music, it was argued, failed to do this. Music without poetry was far too ambiguous in its imitation; it was without definite character or, at this time, sometimes without any character at all. 'Description [in Poetry] runs yet further from the things it represents than Painting,' Joseph Addison wrote early in the century as he contemplated the classification of the evolving fine arts. 'For a picture bears a real resemblance to its original,' he explained, 'which letter and syllables are wholly void of.'

---

[45] See Kristeller, 'The Modern System of the Arts', rpt. in M. Weitz (ed.), *Problems in Aesthetics: An Introductory Book of Readings* (New York/London, 1959), 109; Du Bos, *Réflexions critiques sur la poésie et sur la peinture* (Paris, 1719) quoted in P. Le Huray and J. Day (eds.), *Music and Aesthetics in the Eighteenth and Early-Nineteenth Centuries* (Cambridge, 1981), 21. In the 1740s Charles Batteux established a definitive classification of the fine arts based on the principle of imitation (*Les Beaux Arts réduits à un même principe* (Paris, 1746)).

But then, he added, it would be even 'more strange to represent visible objects by sounds that have no ideas annexed to them and to make something like description in music'.[46] That view prevailed for a long time. And even when imitation was defined as copying the ancients, musicians found their models restricted to vocal music, not just because little if any instrumental music had survived, but because they recognized that the 'civilized' music of Antiquity had taken a form in which the word had primary status.[47]

Instrumental music did not only fail to meet the conditions of imitation. The situation was worse than that. It also failed to embody beauty because of its contrapuntal complexity. Such complexity deprived music of internal order and coherence. In its Baroque style it lacked the beauty so long associated with simple and natural form. 'This noise', Count Pococurante duly informed Candide, 'can give half an hour's amusement; but if it lasts any longer it bores everyone, though no one dares to admit it.' 'Music to-day', the Count concluded, 'is nothing more than the art of performing difficult pieces, and what is merely difficult gives no lasting pleasure.'[48]

Instrumental music was downgraded still further by an old guard who remained convinced that because instrumental music lacked specificity or intelligible and concrete meaning, it failed to teach adequately a moral or political lesson. The fine arts, independently classified or not, were still required to carry a specific moral significance. Theorists were quite content, then, to continue supporting the production of word-based music. Whatever order, coherence, or credibility

---

[46] Addison, *The Spectator*, 27 June 1712, ed. G. A. Aitken (London, n.d.), 58. In the early 18th cent. it was held that if music was to imitate many different affects (passions) or just a single one, the imitation still had to be clear and distinct. Many theorists also accorded poetry a higher status than Addison chose to give it. In the process of determining the concept of a fine art, considerable attention was given to the relations between the various arts, and such is discussed in great detail by B. Hosler, *Changing Aesthetic Views of Instrumental Music in Eighteenth Century Germany* (Ann Arbor, 1981); Kristeller, 'The Modern System'; Neubauer, *The Emancipation of Music*; and Winn, *Unsuspected Eloquence*.

[47] Cf. Burney's comment (*Musical Tour*, 19): 'Poetry, painting, and sculpture have had their rise and declension; have sunk into barbarism; have emerged from it . . . and mounted to a certain degree of perfection, from which they have gradually and insensibly sunk again to the lowest state of depravity: and yet these arts have a standard in the remains of Antiquity, which music cannot boast. There are classics in poetry . . . which every modern strives to imitate; and he is thought most to excel, who comes nearest to those models.'

[48] Voltaire, *Candide*, tr. J. Butt (Harmondsworth, Middx., 1947), 118. Contrary to Voltaire's observation, it seems that many people were willing to admit their dislike of instrumental music.

melodic composition could have, it had to be derived from its traditional role as accompaniment or 'mistress' either to the poetic form or the religious text. 'In the union of poetry and music', James Harris argued in 1744, poetry must 'ever have the precedence, its utility as well as dignity being by far the more considerable'.[49]

The theorists were not alone. Many composers chose to adopt the theoretically acceptable conclusion of their day. A good musician, Rameau firmly concluded in 1722,

should surrender himself to all the characters he wishes to portray. Like a skillful actor he should take the place of the speaker, believe himself to be at the locations where the different events he wishes to depict occur . . . he must declaim the text well . . . so that he may shape his melody, harmony, modulation, and movement accordingly.[50]

But that other practising musicians were deeply discontent with the lowly status awarded to instrumental music was being expressed with ever-greater vigour—even if to no immediate effect. Their discontent and subsequent 'revolutionary' action was indirectly revealed as early as 1720 in Marcello's satire on the modern theatre. Marcello described most modern composers as no longer wanting or feeling it necessary to have knowledge of 'the rules of good composition'—practice and a few general principles seemed to suffice. Nor did they find occasion any more for acquaintance with poetry, he said. In fact, they resist reading the poem before setting it to music, for fear of over-loading their imagination and oppressing their genius. They only produce airs that are accompanied by the whole orchestra. In order to compose in the modern taste, Marcello bemoaned, it now appears indispensable, above all things, for musicians to make plenty of noise.[51]

It seemed that composers of instrumental music were in a desperate situation. And that situation seemed only to worsen when theorists added a new complaint to their already lengthy list. They proclaimed

---

[49] *Three Treatises: The First concerning Art, the Second concerning Music, Painting, and Poetry, the Third concerning Happiness* (London, 1744), ch. 6, § iii (quoted in Le Huray and Day (eds.), *Music and Aesthetics*, 39).

[50] *Treatise on Harmony*, 156.

[51] Recorded in Hogarth, *Music History*, 71, and Strunk, *Source Readings*, 525 f. According to Strunk, the essay marked 'a spontaneous reaction [in Italy] of those concerned with the musical theatre as a temple of dramatic art against all those . . . who wanted to use it for the gratification of their personal vanities'. One should also recall here the vehement arguments against modern theatre expressed by Jean-Jacques Rousseau in a letter written in 1758 to J. d'Alembert (*Politics and the Arts*, 'Letter to d'Alembert on the Theatre', tr. A. Bloom (Ithaca, 1960)).

that musical complexity had the unfavourable consequence of distracting the listener from the real purpose of art, not only when that purpose was understood as the fulfilment of a religious or pedagogical function, but also when it was understood as the production of pleasures triggered by the imagination. Only such pleasures could and would be lasting, and only lasting pleasures were of value. At least this particular complaint was not against instrumental music *per se*, but only against its more complex examples.[52]

The emphasis on imagination received significant expression in the work of Alexander Gottlieb Baumgarten. In the 1750s, he used the term 'aesthetics' for the first time to refer to the scientific study of the formal, deductive, and a priori principles of sensory cognition. He recognized, however, that aesthetic perception, unlike most other forms of perception, involved a non-rational, non-cognitive element. He thereby broke with the Cartesian tradition and resisted the particular and still popular belief that 'the artist's brush', as his contemporary and student, Winckelmann put it, should like Aristotle's pen be dipped in reason. Winckelmann had ended his *Reflections* with this very thought. Though Winckelmann had an enormous influence on the future of the arts, it was Baumgarten who helped give the arts a content and meaning that made it suitable for later theorists to speak of art's, or rather the artist's, irrational character. Speaking positively (though ironically, perhaps) of such irrationality as a 'negative capability', John Keats was later to describe the ability of artists to be 'in uncertainties, mysteries, doubts, without any irritable reaching after fact and reason'.[53]

Towards the late eighteenth century, it was expected less and less that art convey some explicit moral, religious, or rational meaning, at least on the surface. Interest in the imaginative faculty was leading instead to new concepts of the Beautiful and Sublime, concepts that would cut off the artistic from the scientific and the moral. While the Beautiful (and, for some, the Sublime) were to be intuited by imagination in a pure aesthetic experience, the True and the Good were to be

---

[52] The idea that the arts are associated with pleasures of the imagination gained much currency during the 18th cent., though its expression never departed very far from Addison's early one in 1712. The pleasures of the imagination, Addison wrote, 'proceeds from that action of the mind which compares the ideas arising from the original objects, with the ideas we receive from the statue, picture, description or sound that represents them' (*The Spectator*, 27 June, 59). As we have already seen, however, Addison had little faith in sound's ability to represent ideas with any clarity.

[53] *Letters of John Keats*, ed. M. B. Forman (Oxford, 1952), Letter 32 (1817), 71.

grasped by distinct mental faculties. Was it possible that instrumental music could be beautiful independently of its being good? Until these distinctions became effective, however, and until they were taken to an extreme by romantic theorists, instrumental music, even regarded as an art capable under certain conditions of arousing pleasures of the imagination, was accorded its generally low status.

Instrumental music, in failing every demand ever put forward by those wishing a valuable end for music, came to be described as '*unverständlicher Mischmasch*'. That was the final insult. It was a skill, theorists said, suitable for persons interested only in 'somersaults', 'rope-dancing', and other such acrobatic feats, for persons, in other words, evidently uninterested in serious pursuits. But the question first asked at the turn to the sevententh century by Bernard Le Bovier de Fontenelle (*secrétaire* of the French *Royal Academy of Sciences*) continued to nag the theorists. 'Sonata,' he had asked, 'what do you want of me?' Only those optimistic about the value of instrumental music had an answer to justify its production. For the rest, such music seemed to want of them something they were not prepared to give.[54]

By the end of the eighteenth century, such philosophical malcontents as Friedrich von Schlegel had begun urgently to seek for instrumental music an acceptable status. 'Must pure instrumental music not create its own text?', he asked.[55] The search was not this time in vain. But its success depended on a radical change in aesthetic attitude, one that transformed the classical into the romantic age. The change generated an understanding of music that had not been seen before, or if seen had not been allowed to carry any theoretical weight. It ended up founding a 'specifically musical' music and a very 'civilized' understanding thereof. It was an understanding immediately intelligible to theorists willing, first, to abandon the belief that music should serve an extra-musical, religious, or social end, and then to adopt, in its place, the belief that instrumental music could be a fine and respectable art in service to nothing but itself.

---

[54] I have followed Hosler's argument here (*Changing Aesthetic Views of Instrumental Music*, esp. 26 and 42). Cf. also Neubauer, *The Emancipation of Music*, 67, and *Roger North on Music*, 129, on rope-dancing.

[55] 'Muss die reine Instrumentalmusik sich nicht selbst einen Text erschaffen?', 'Athenäums-Fragmente', *Kritische und theoretische Schriften* (Stuttgart, 1978), 140.

6

Musical Meaning:
Romantic Transcendence and the
Separability Principle

> When we speak of music as an independent art should we not
> always restrict our meaning to instrumental music, which,
> scorning every aid, every admixture of another art . . . gives pure
> expression to music's specific nature, recognizable in this form
> alone? It is the most romantic of all the arts—one might almost
> say, the only genuinely romantic one—for its sole subject is the
> Infinite.
>
> (E. T. A. Hoffmann,1813[1])

There is a twist in the tale as to how music achieved its emancipation
from the extra-musical and how it simultaneously found its new
emphasis to be placed on works. Though for centuries musicians sought
a status specifically for instrumental music dictated by external—
religious and moral—principles, so that this music could be allied to
other practices intrinsically embodying those principles, it achieved
its respectable status only when it became emancipated.

But emancipation is often a double-sided process. Music, like many
emancipated persons or races, first sought its independence by becoming
like something else that was or seemed already to be emancipated.
Music allied itself to the fine arts, notably the plastic arts of painting
and sculpture, arts that had found some (if not complete) independence during the eighteenth century. Yet while music endeavoured to
become another art, the very fact of being a fine art (and the very fact
of being emancipated) dictated that music be conceived and assessed
on its own terms. Music as an emancipated fine art was ideally and
gradually to become, then, an independent, autonomous practice,
depending on nothing ultimately but itself—its own internal ideals
and its own medium—for its functioning, power, and significance.
What kind of aesthetic theory could sustain this bid for emancipation?

---

[1] 'Beethoven's Instrumentalmusik': I have used Strunk's translation here (*Source
Readings*, 755).

I

The first step in the emancipation of music from the 'extra-musical', and, importantly, the corresponding emergence of its work-concept, depended on the fusion of two traditional concepts: music and productive art.

Until the mid-eighteenth century, music and productive art functioned in relative, though diminishing, independence from one another. Originally the term 'art' (*ars* or *technē*) had wide designation. It designated a skill in making products, a skill in practical performance, and a skill in theoretical activities of the mind. For our purposes the first two moments are the important ones. Thus, for many arts— sculpture and painting—the emphasis was on making a concrete product. For other arts—notably music—emphasis was placed on the skilled doing, the performance as such, though of course the making of instruments or theoretical speculation about music put the emphasis elsewhere.

Because the productive or mechanical arts emphasized the physical end-product of the activity—the concrete object made—one finds numerous references in the literature to works of art. The works were the products. Still, these references are apt to mislead. Before the late eighteenth century, the productive arts, like performance or theoretical arts, were usually also conceived functionally, in the service of social, political, and religious ends. Even in early formulations of fine art, extra-artistic factors crept in. Though 'aesthetic' features had increasingly come to characterize the plastic arts since the Renaissance, one could none the less still speak, even in the 1750s, of a person painting pictures or sculpting models for some purpose, just as one could speak of the purpose fulfilled when someone sings, plays an instrument, or recites poetry.

Nowadays, because of the prevalence of a functionless and autonomous aesthetic, a rigid conceptual and evaluative distance is imposed between creative activity, the product of that activity, and the function of that activity (if there is one), such that we find no difficulty in distinguishing the three aspects. This was less the case prior to the late eighteenth century. Before then, the idea of something's being done or made was effectively inseparable from the idea of its having been done or made in fulfilment of a designated function. The idea of producing works was not usually considered an end in itself. Of course, artworks could still be appreciated as significant contributions to to art, even if their production was motivated by extra-artistic functions.[2]

Before the emergence of fine art in the eighteenth century, the class of works of art included not just works imbued with 'aesthetic' value (or something equivalent) but all utilitarian products of skilled or mechanical labour. Works of art were not sharply distinguished, in modern parlance, from craft-products. Art production was guided as often by needs fulfilled by simple manual skill and labour, or by needs of a given market, as by 'higher' principles and functions. To have mastered a craft was to have mastered the rules of a particular form of material production and to have produced a good or useful work of art: the making of a vase, for example.

When theorists spoke of art with reference less to concrete products than to skills in making, doing, and thinking, the application of the term 'art' became wider than that of 'work of art'. For by referring to skill, the term 'art' could designate both skills that resulted in a product and those that did not. Hence the performance 'upon an instrument' and the composing of music could be referred to as skills—as arts—even though they resulted in 'ephemeral' events rather than concrete or lasting products. Certain arts yielded works of art, others did not. No one better demonstrated the difference than St Basil, though for him the distinction was not evaluatively neutral. Writing in the fourth century CE, he informs us that

Of the arts necessary to life which furnish a concrete result there is carpentry, which produces the chair, architecture, the house; shipbuilding, the ship; tailoring, the garment; forging, the blade. Of useless [or 'destructive'] arts there is harp playing, dancing, flute playing, of which, when the operation ceases, the result disappears with it.[3]

---

[2] Cf. Raymond Williams's discussion of the transition from functional, performative, and adjectival uses to substantive uses of terms relating to the arts in his *Culture and Society, 1780–1950* (Harmondsworth, Middx., 1963), ch. 2; also Patricia Carpenter's suggestion that such a transition in music was noted in 1732 by Johann Gottfried Walther in his *Musicalisches Lexicon* (The Musical Object', ed. C. Seltzer, *Current Musicology*, Special Project, 5 (1967), 59). The passage Carpenter refers to is Walter's entry for '*Musica*': '(*lat. ital.*) Musique (gall.) . . . wird als ein Adjectivum durchgängig Substantivè gebraucht, und bedeutet überhaupt die Ton-Kunst, d. i. die Wissenschafft wohl zu singen, zu spielen, und zu componiren.' Carpenter uses this as evidence for the presence of a work-concept, though as far as I can tell the passage only indicates a move from describing something as musical to describing something as music.

[3] Quoted in P. Weiss and R. Taruskin, *Music in the Western World: A History of Documents* (New York, 1984), 27.

So long as music was conceived as a 'performance' rather than as a productive art, it was not generally understood as involving the production of works. Music as an art of skilled performance did not result in lasting or concrete products. Music, associated with its own specific functions, imbued with certain clearly specified religious and theoretical meanings, was effectively excluded from the category of the productive arts or crafts. Thus the lack of reference in the early literature to works specifically of music.

So conceived, music could survive without producing works. Functional performances sufficed. But as we already know, they did not suffice forever. Circumstances gradually changed so that music came to be regarded as an art that resulted from the activity of composition not just in performances but also in works of art. A major step in this development was marked by Listenius in the sixteenth century, only if we interpret his words as expressing the need for the composition or making of music to result in something tangible that survives after the music has, as it were, been played out in performance. In any event, it was a long time after the sixteenth century before such a need was completely fulfilled.

II

Traditionally, arts yielding works of art were characterized in a way that would later be appropriate for characterizing works of music. Following Aristotle, Seneca characterizes productive art by identifying 'five causes': 'the material, the efficient, the formal, the archetypal and the final'. Of the statue, he explains, 'the bronze is the material cause, the sculptor the efficient, the shape given to it the formal, the model copied by the maker the archetypal, the end held in view by the maker the final, the statue itself, the resultant of these several causes'.[4] Further formulations of compositional/creative and artefactual conditions can be found in the literature from Aristotle to Aquinas, as can references to the necessary personalization of each work of art. Often one hears, for example, that the aim of productive arts is to bring something into existence, and, though the rules of each art are universal, what one brings into existence is individual. Many of these conditions readers should in fact recognize, since they were formulated also in the guise of identity conditions by the analytic theorists.

[4] Seneca, *La lettera 65 di Seneca*, ed. G. Scarpat (Brescia, 1970), 14–16. (I have used Tatarkiewicz's translation, *History of Aesthetics*, i. 300.)

With the emergence of fine art, a distinction crystallized between art and craft. It rested on a distinction being drawn between aesthetic value and functional utility. Craft products were good because of their function in the everyday world. Art was beautiful because, among other things, and as it would soon be expressed by romantic theorists, it could transport us to higher, aesthetic realms. During the aestheticization of fine art, starting early, though peaking around 1800, the understanding of an artwork significantly altered. Conditions previously associated with productive art, especially those of creativity, product, artefactuality, and perseverance, were given a new aesthetic significance such that their use in the domain of fine art would differ—especially in matters of value—from their use in the domain of craft.

As music began to be understood first and foremost as one of the fine arts, it began clearly to articulate its need for enduring products—artefacts comparable to other works of fine art. Hence the emergence of a work-concept in the field of music in the mid- to late-eighteenth century. As the work-concept formed, emphasis began to be put on those very same conditions associated with the other fine arts, conditions previously underemphasized, or overridden by others, in the history of music. In this respect, then, the origins of the concept of a work of music are to be found less in the history of music before the 1750s than in the history of the productive arts.

But it was only with the romanticization of fine art around 1800 that theorists found a really successful way to give substance to the idea of a musical product. At this moment, the work-concept became the focal point, serving as the motivation and goal of musical theory and practice. All references to occasion, activity, function, or effect were subordinated to references to the product—the musical work itself. What, we must find out now, did the romantic aesthetic contribute towards this process of turning music first into a productive art, and then into an art that stressed the product above all else?

III

Until the late eighteenth century, as the previous chapter revealed, theorists found a particular way to give music meaning. They attributed to it specific 'extra-musical' meanings to render it a worthy contribution to a moral, rational, and religiously upright society. Music's meaning had come from 'outside' of itself, deriving as it did from music's ability to influence and empower a person's religious and moral beliefs, or from its ability to imitate the nature of persons and the world. Theorists argued further that music achieved its external significance when it employed the medium of words. Words rather than pure melodies were intelligible. They had concrete and specific semantic content and produced similarly concrete effects. Even when theorists claimed that the aesthetic ideal should express something general, they still thought that this end was achieved through concrete or specific, representational means. A direct consequence of all these claims was the near-complete rejection of purely instrumental music as a worthwhile form of musical production. Yet the very idea that instrumental music lacked both referential significance as well as concrete and specific content, the very idea, in other words, that had led to the rejection of such music as unworthy, turned out to be the key to finding for this music its long-sought-after respectability.

Under the new aesthetic, here broadly and collectively being called the romantic aesthetic, many new doctrines of a more or less romantic, formalist, and idealist inclination were proposed. In several of these, despite the particular inclination, a basic argument was put forward. It was a very complicated argument, however, for it rested upon an interplay between two claims which we nowadays separate more sharply than theorists originally did. The first claim concerns the *transcendent* move from the worldly and particular to the spiritual and universal; the second concerns the *formalist* move which brought meaning from music's outside into its inside.

The argument began with theorists reconceiving the fine arts in such a way that music came more closely than ever before to be allied to the other fine arts, and continued with theorists coming to see instrumental music as exemplary of such art. First, theorists claimed, the significance of fine art lies not in its service to particularized goals of a moral or religious sort, or in its ability to inspire particular feelings or to imitate worldly phenomena. It lies, rather, in its ability to probe and reveal the higher world of universal, eternal truth. This ability originates, according to Gustav Schilling, in 'man's attempt to transcend the sphere of cognition, to experience higher, more spiritual things, and to sense the presence of the ineffable'. Theorists then argued that instrumental music, without particularized content, is the most plausible candidate for being the 'universal language of art'. Such music provides a direct path to the experience of a kind of truth that transcends particular natural contingencies and transitory human feelings. Schilling made the point again: 'No aesthetic material is better suited to the expression of the ineffable than is sound.'[5]

Many theorists suggested, in fact, that instrumental music was now the most pleasurable and beautiful of the arts. It was the art whose beauty was grasped most immediately and readily in the pure aesthetic intuition of those with high cultivated taste. 'In instrumental music,' Ludwig Tieck wrote, 'art is independent and free; here art phantasizes playfully and purposelessly, and nevertheless art attains the ultimate.'[6] As music came to be heard as a melodious 'song without words', it was quickly attributed a status which, a few years earlier, one would not have imagined was possible. For many it had become the finest of the arts, to which all the others had now to aspire. 'Music, you I praise above all,' Grillparzer waxed poetical: 'To you the highest prizes fall, | Of the three sister arts | You the truest, uniquely free.'[7] Walter Pater's famous proclamation was to give explicit articulation to an idea that at the time of utterance was already at least half a century old. 'All art', he wrote in the 1870s, 'constantly aspires to the condition of music.'[8]

The suggestion that music carried transcendent meaning led soon enough to the view that instrumental music did more than point to the transcendent. It also embodied it. This claim made sense precisely for the reasons which had been used to reject instrumental music in earlier times. The lack of intermediary, concrete, literary or visual content made it possible for instrumental music to rise above the status of a medium to actually embody and become a higher truth. 'It is not . . . by imitation properly, that instrumental music [engages our attention]', Adam Smith argued in 1795:

instrumental Music does not imitate, as vocal music, as Painting, or as Dancing would imitate, a gay . . . or a melancholy person; . . . it becomes itself a gay . . . or a melancholy object; . . . What is called the subject of [instrumental] Music . . . is altogether different from what is called the subject of a poem or a picture, which is always something which is not either in the poem or in the picture.[9]

[5] Le Huray and Day (eds.), *Music and Aesthetics*, 470.

[6] Quoted in Hosler, *Changing Aesthetic Views of Instrumental Music*, 190.

[7] 'Tonkunst dich preis ich vor allen | höchstes Los ist dir gefallen | Aus der Schwesterkünste drei | Du die freiste, einzig frei.' 'In Moscheles Stammbuch', 10 Oct. 1826, *Sämtliche Werke: Gedichte*, iii.1 (Vienna, 1937), 30. The three sister arts are music, poetry, and dance.

[8] W. Pater, 'The School of Giorgione', *The Renaissance: Studies in Art and Poetry*, ed. A. Phillips (Oxford, 1986), 86.

[9] *The Works of Adam Smith* (5 vols.; London, 1811), v. 287–301. Smith's writings mark a transitionary phrase. Whilst he argues against a mimetic theory for instrumental

It was the shift from imitation of particulars to immediate expression and embodiment of the transcendent that ultimately gave to instrumental music its new meaning. Indeterminate on a concrete level, it was deemed utterly meaningful on a transcendent one. Precisely in its indeterminacy was it able to capture the very essence of emotion, soul, humanity, and nature in their most general forms.

So far, music had achieved a two-pronged emancipation: first from its service to particularized, extra-musical goals; second from its dependence on words. Words, unable to transcend semantic specificity or particular cognitive content, could not constitute a universal medium as successfully as pure sound. 'Music has developed into a self-sufficient art, *sui-generis*, dispensing with words,' Herder thus proclaimed in 1800. But, he reminded his readers, 'the slow progress of music's history' demonstrates just how hard it has been for music 'to cut herself free from her sisters—mime and the word . . . to establish herself as an independent art.'[10]

Music's emancipation from the word turned out to be far less problematic than its emancipation from the extra-musical, however. For the move towards transcendence had proved itself insufficient as yet to give music the purely musical meaning it desired. In other words, though freed of the constraints of social functions determined by church and court, though freed from service to a text, the transcendent move had not freed music of its obligation to be meaningful in extra-musical, spiritual, and metaphysical ways. Formalists subsequently came forward to provide the necessary, next step. Music, they argued, is intelligible not because it refers to something outside of itself, but because it has an internal, structural coherence. It consists in an internal and dynamic stream of purely musical elements, in Hegel's terms, in an 'abstract interiority of pure sound'.[11]

music, he still maintains that other musical forms, specifically opera and vocal music, function mimetically through their alliance with the word.

[10] Le Huray and Day (eds.), *Music and Aesthetics*, 257.

[11] The idea of an 'abstract interiority (*formellere Innerlichkeit*) of pure sound' is found in Hegel's discussion of music in his *Aesthetics: Lectures on Fine Art*, tr. T. M. Knox (2 vols.; Oxford, 1975). I have used Stephen Bungay's translation of this phrase (*Beauty and Truth: A Study of Hegel's Aesthetics* (Oxford, 1984), 135). Note that formalists increasingly rejected the idea that music embodied indefinite or transcendent content of any extra-musical sort. The content of music, they argued, resides only in the tonal relations of music and nothing else; the content of music is just the tonally moving forms (*tönend bewegte Formen*). This view gradually paved the way to the view that a language for describing music, such as one using emotive or expressive predicates, can only ever be metaphorical in status, since the content of instrumental music can never

Was there a contradiction here? How could the formalist demand that music 'mean itself' be reconciled with the demand that music have spiritual and metaphysical meaning? Friedrich von Schelling provided the answer.

Music brings before us in rhythm and harmony, the [Platonic] form of the motions of physical bodies. It is . . . pure form, liberated from any object or from matter. To this extent, music is the art that is least limited by physical considerations in that it represents *pure* motion as such, abstracted from any other object and borne on invisible, almost spiritual wings.[12]

Schelling achieved reconciliation by using the notion of pure form in two ways: to show how music, of all arts, could be the universal and spiritual language and to show how music could have purely musical meaning. His reconciliation took place, however, against the background of a more complex theoretical transition in aesthetic and conceptual theory.

The transition falls under four, closely related descriptions: first, there was a move from seeing music as having value and meaning solely by virtue of its service to another thing, to seeing it as having value and meaning in itself (even though similar values and meanings happened also to be found elsewhere). Second, the 'extra-musical' obligation for music to resonate with our spiritual lives was now believed to belong to the musical domain. It had become a musical obligation. The purely musical, in these terms, was now synonymous with the moral, the spiritual, and the infinite in its uniquely musical form. Third, matters in relevant circumstances considered extra-musical could in other circumstances be regarded as purely musical. It was a case of notions and values being appropriated by musical theorists from 'the outside' to fill out their new concept of the purely musical. Fourth, the distinction between the musical and the extra-musical was allowed to function on a worldly but not on a transcendent level. Functioning on a level in which worldly principles of individuation no longer apply enabled one to accept a double-sided view of musical meaning, that it be transcendent, embodied spirituality and purely musical at the same time.[13] In sum, the new romantic aesthetic

be reduced or captured in a non-musical language. For more on this, see Eduard Hanslick's *On the Musically Beautiful*, tr. and ed. G. Payzant (Indianapolis, 1986), 9, 29, and 32.

[12] Le Huray and Day (eds.), *Music and Aesthetics*, 280.

[13] The process by which music becomes defined in terms of the musical might also be described as follows: as a series of 'successive negations' of those extra-musical

allowed music to mean its purely musical self at the same time that it meant everything else.

The transition just described had to explain many things. Before all else, it had to explain the precise sense in which religious and moral attitudes could continue to exert their influence on the character of music without that influence threatening music's new-found autonomy. Herder, for example, addressed this matter when he noticed that 'one single compelling force' had been needed to give music her 'independence and to free her from alien help'. What was that force? he had asked. His answer was that it was 'religious awe'. But Herder quickly realized that the influence of religion was now quite obscure if not mystical, for when he spoke of religious awe, he was not thinking of forms of religious worship taking place in the church, for which music had once served as an aid and accompaniment. He was thinking, rather, of some transcendent, 'religious' impulse. Indeed, as many theorists were realizing quite explicitly, the development of the notion of fine art was depending upon the cessation of a religiously based society. Many modes of behaviour traditionally associated with church worship were being adopted by the institution of fine art, as being utterly appropriate in the treatment and appreciation of the new works.[14]

IV

Both the emancipation of instrumental music and the romanticization of fine art depended upon the use of a single principle—a principle with roots in religious doctrine. I shall henceforth refer to it as the *separability principle*. At the end of the eighteenth century, it became the custom to speak of the arts as separated completely from the world of the ordinary, mundane, and everyday. The essence of fine art comprised the basic idea of severance from anything associated with the transient, contingent world of mere mortals. 'Whatever is familiar, or in any way reminds us of what we see and hear every day,' Sir

meanings music formerly had, assigning the remainder to the musical. This description, while accurate, does not go far enough, since it does not account for the appropriation of the extra-musical that was also involved in forming the concept of the musical. For reference to 'successive negations', see P. Bourdieu's *Distinction: A Social Critique of the Judgement of Taste*, tr. R. Nice (Cambridge, Mass., 1984), 56.

[14] Herder quoted in Le Huray and Day (eds.), *Music and Aesthetics*, 257. On the relation between religion and art, see e.g. Lessing's paradigmatic statement in his *Laocoön* (tr. H. B. Nisbet, in Nisbet (ed.), *German Aesthetic and Literary Criticism: Winckelmann, Lessing, Hamann, Herder, Schiller and Goethe* (Cambridge, 1985), 87).

Joshua Reynolds surmised in 1798, 'perhaps does not belong to the higher privinces of art, either in poetry or painting.' Hegel was less tentative in his use of the separability principle. A work of fine art, he wrote 'cuts itself free from any servitude in order to raise itself to the truth which it fulfils independently and conformably with its own ends alone. In this freedom is fine art truly art.'[15]

It was just this idea of separability that set up classificatory boundaries, evaluative criteria, and normative principles that were, for the first time, really in the interest of the arts and artists themselves. It set up these boundaries with the result, also, that music as the most abstract of all the arts became the paradigm art in many, even if not in all, considerations of the new aesthetic.[16]

Yet the separability of fine art from the world of the everyday cannot now, and could not then, be taken too literally. It had to reveal itself and to find its stability through a complex array of metaphorical beliefs. Many of these shared a peculiar feature, which might well be called the *romantic illusion*. It captured the ability of an object, a person, or an experience, to exhibit simultaneously the character of the human and of the divine, of the concrete and of the transcendent. 'Inasmuch as I give', Novalis wrote, 'the lowly a higher meaning, the common a hidden aspect, the known the dignity of the unknown, the finite an infinite appearance, thus I romanticize them.' Often the illusion was expressed as if rendering an oxymoron or, as one might describe it, a Janus-like gaze. Familiar examples are Kantian phrases such as 'purposiveness without purpose' and 'disinterested attention'. Phrases like these, as we shall see, not only presupposed understanding of the separability principle, but also pointed usefully to the opposition of a new aesthetic notion to its predecessor.[17]

[15] Reynolds, *Discourses on Art* (New York, 1961), XIII. 207; Hegel, *Aesthetics*, 7.

[16] M. H. Abrams speaks of a 'Transcendental Ideal' in *The Mirror and the Lamp: Romantic Theory and the Critical Tradition* (Oxford, 1953), 42 ff. For related discussions of the 'process of aestheticization', with rather different emphases, see Gadamer's *Truth and Method*, pt. I; T. Eagleton, *The Ideology of the Aesthetic* (Oxford, 1990), chs. 2 and 3; and Bourdieu, *Distinction*, pt. I.

[17] Novalis quoted in Schafer, *E. T. A. Hoffmann and Music*, 42. The 'romantic illusion' has origins in German Mysticism, and earlier discussions of disinterestedness and taste are to be found in treatises by Hume and Shaftesbury. In both the latter, there was a tendency to synthesize moral and 'aesthetic' principles. To be sure, later theorists also found reason to connect aesthetic and moral issues. Kant's 'sensus communis' is closely connected to the ideal of a cultivated person of good morals (*Critique of Judgement*, tr. J. C. Meredith (Oxford, 1952), § 20). Schiller spoke of the 'tragic sublimity' of the aesthetic state as one sought for oneself an harmonious and integrated personality ('On Naive and Sentimental Poetry', in Nisbet (ed.), *German Aesthetic and*

The separability principle was abundantly used by theorists around 1800. It was used to redefine various major distinctions: those between art and nature, art and craft, and the civilized and the popular; various aesthetic notions: artistic form, content, and medium; and various activities: the creation and reception of fine art. Each use of the principle—as the rest of this chapter demonstrates—ended up presupposing, demanding, or referring indirectly to one thing: the existence of an artwork at the centre upon which one's aesthetic concerns could focus.

V

With the predominance of the mimetic principle in the early eighteenth century, the significance of the new fine arts had derived from their external relation to nature. With the use of the separability principle in later years, this external relation was cut. A new and perhaps even more complex relation between art and nature was put in its place.

Initially, the primary function of fine art was to imitate nature as well as possible. Success in this would render the imitation valuable. That art was conceived to be in the service of nature stemmed from the long-standing belief that nature manifested unsurpassed beauty. Art contrarily obtained its worth or beauty through its mimetic and therefore externalized connection to nature. It was judged relative to its degree of naturalism. Cicero's early dictum that 'in all cases reality undoubtedly surpasses imitation' was confirmed in the 1730s by Scheibe. 'Art must imitate nature,' he wrote:

So soon as this imitation is exceeded Art is condemned for being in disagreement with nature. Art does not endow nature with beauty but nature endows art. Nature possesses by itself everything meritorious and need not borrow any rouge from art. The greater, or rather, the more extravagant art is [as, in his opinion, J. S. Bach's music had become], and the farther it goes its own way, the more does it alienate itself from nature, and the less will one be able to arrive at good taste. . . . It is therefore a fact that too much art obscures true beauty.[18]

Literary Criticism, 177–232). Goethe also spoke of artistic pursuits in connection with self-cultivation [*Bildung*] (Nisbet (ed.), *German Aesthetic and Literary Criticism*, 233–58). However, there was also a strong recognition by these later theorists of the self-sufficiency and autonomy of artistic/aesthetic activities, not shared by earlier theorists.

[18] I. Willheim, 'Johann Adolf Scheibe: German Musical Thought in Transition', diss. (Univ. of Illinois, 1963), 104. In Ch. 1 Sect. ix, I mentioned Scheibe's opposition to modern developments in Bach's composition.

The romantic aesthetic turned the table around. First a notion of human spirit was incorporated into the grand concept of Nature. Alternatively, some theorists replaced Nature altogether with Human Spirit. In either case it was then argued that the arts, with their special powers of abstraction, could reveal the essence and generality of Nature and/or Spirit more successfully than those natural phenomena found in the physical world. Aesthetic value or beauty was no longer to be sought first in nature and then in art, but vice versa. Hegel spoke of art as being of a higher rank than nature since 'no natural being is able, as art is, to express the Divine Ideal'. Johann Adam Hiller illustrated the turn-around more soberly when he wrote specifically about music that it 'is not so much an imitation of the song of the passions and of the heart, as it is an artful combination of tones arranged according to the nature of the instruments that are played and judged as correct more in accordance with art than with nature'.[19]

It would be erroneous to think naturalism was given up, however. It simply took on a new meaning. The value of an artwork was judged by its success in exhibiting simultaneously 'high' artistic content and an aura of naturalness. The aura of naturalness depended upon one's recognizing the intimate relation between two aspects of the artistic: one, the element of human making, the other, the transcendent end to which the making was directed. To obfuscate matters, this end was conceived either as Spirit or as Nature, or synthetically as both. The impact of these claims is perhaps clearer than the claims themselves for they gave rise to what became a familiar illusion of naturalness. This illusion entailed either that art and nature existed as a synthetic unity on a transcendent level or, on a lower level, that human making was truly fine when the product of the making looked as if it had *not* been made by human hands. Goethe immediately recognized the illusion as he wrote that 'the artist has a two-fold relation to nature; he is at once her master and her slave.'[20]

To give credence to the illusion, theorists argued that a work of fine art is created by an artist—a genius—who in following the rules of art is not constrained by their inherent limitations (the limitations, say, of

---

[19] Hegel, *Aesthetics*, 29; Hiller quoted in Hosler, *Changing Aesthetic Views of Instrumental Music*, 133. An earlier version of this view was articulated by Johann Nicolaus Forkel (renowned as the first biographer of J. S. Bach). 'The man of genius does not distance himself from nature,' he writes, 'what he does is penetrate more deeply into her mysteries, enticing her more recondite secrets from her and revealing them to others' (Le Huray and Day (eds.), *Music and Aesthetics*, 178).

[20] Quoted in Beardsley, *Aesthetics*, 260.

technical knowledge). 'What can be carried out' with 'rule-providing theories', Hegel wrote,

can only be something formally regular and mechanical. Being abstract in content . . . rules reveal themselves as wholly inadequate, in their pretence of adequacy, to fill the consciousness of the artist, since artistic production is not a formal activity carried out in accordance with given specifications.[21]

When Berlioz proclaimed music to be 'in the flush of its youth' and 'emancipated', he too described this phenomenon in terms of a liberation from rules.

Music does what it wants. Many of the old rules are no longer binding: they were made by inattentive observers or by mundane spirits for other mundane spirits. . . . Many forms have been used too much to be used again.[22]

By transcending rules, artists were able to meet the demand that they create works that are experienced *as if* they were free from artificial making—as if they were natural. Kant thus argued that 'a product of fine art must be recognized to be art and not nature. Nevertheless the finality of its form must appear . . . as if it were a product of mere nature.' Forkel found that the illusion of naturalness had been successfully created by J. S. Bach, whom he said composed so easily and naturally 'that the workmanship is not perceptible and the composition sounds so smoothly as though it were in the free style'.[23]

From the prescription for naturalness it followed, furthermore, that a work of fine art had to conceal its human origins or its moment of creation. Once created it should exhibit a permanence and self-sufficiency that would separate it from all worldly or historical contingency. Each work thus required a kind of false consciousness on the part of the public for it to survive and function in the way conceived for it. Creative artists would have to demonstrate their independence from their creations by interfering with and destroying the impression that their works were created by human hands, and in

---

[21] Hegel, *Aesthetics*, 25–6.
[22] Hector Berlioz, 'Concert de Richard Wagner: La Musique de l'avenir', *A travers chant: Études musicales: Adorations, boutades et critiques*, 2nd edn. (Paris, 1872), 312: 'La musique, aujourd'hui dans la force de sa jeunesse, est émancipée, libre; elle fait ce qu'elle veut. Beaucoup de vieilles règles n'ont plus cours; elles furent faites par des observateurs inattentifs ou par des esprits routiniers, pour d'autres esprits routiniers. . . . Diverses formes sont par trop usées pour être encore admises.'
[23] Kant, *Critique of Judgement*, § 45, 166; J. N. Forkel, *Johann Sebastian Bach: His Life, Art and Work* (1802), tr. C. S. Terry (London, 1920), 88.

part they could do this by refusing to follow any pre-established set of mundane rules.

Despite the force of the ideal of naturalness, artists were still to be conceived as genuine masters of their art. This demand depended upon the relation between artists and their artworks being seen in a new way. The relation, once subordinated to the external relation between an artwork and that which is imitated, was now formulated as an internal relation of *expression*, more accurately *self-expression*, that bound an artwork to its creative source.[24]

But had not a paradox been generated? How could an artwork be viewed as an embodied expression of an artist as well as an independently existing work that, once created, had meaning without reference to its creator? During the nineteenth and into the present century, a tension evolved between expressive and intentionalist views of art, on the one hand, and formalist and organicist views, on the other. Theories presented in the early nineteenth century managed to embrace both views simultaneously, but they did so only by adopting a position later theorists were reluctant to take.

Early romantic theorists argued that it was on the transcendent, universal level of the 'free' genius that artists gave 'fine' content to their works. One way to counteract the belief in the *human* creation of a work was to attribute a God-like existence to the creator. Artists effectively superseded their status as mere mortals to reach an 'aesthetic state', in Schiller's terms, so that the content of their works would express not the individual or mundane thoughts of a mere mortal, but universal thoughts of which there can be no personal ownership. In this transcendence, one could retain both the distinction between and the synthesis of artwork and artists simultaneously.

This two-fold aesthetic recognized, in other words, work and artist both as separated and as bound to one another. Works could receive, on the one hand, ample attention in their own right as independent and self-sufficient entities. On the other hand, the ability of artists to reach the level of the universal and to express that universal in their works was sufficient to guarantee them their own personal respect and recognition. Christian Gottfried Körner clarified the new task of the artist with these words:

---

[24] Neubauer argues (following Abrams) that the romantic revolution consisted in a metaphorical shift. 'This aesthetic revolution', he writes 'which replaces mimesis (art in relation to nature) by an expressive theory (art in relation to the artist), is a changeover indicated by the metaphors of mirror and the lamp' (*The Emancipation of Music*, 5).

He must portray the dignity of human nature in the products of his creative imagination. He must raise us to his level from our lowly, circumscribed state of dependence and represent to us the Infinite, an Infinite that can otherwise come to us only by Intuition [*Anschauung*].[25]

Why did artists have this task? Just because they had the ability to create embodiments of the Infinite; because they could, unlike 'mere mortals', actually produce fine works of art.

VI

The new complex relation between art and nature presupposed use of the separability principle in other ways as well. Novel descriptions of art's form and content—especially music's—amply demonstrate such use.

Traditionally, theorists had imposed upon musical material the requirement that it sound as natural as possible; for acceptability derived from truthfulness and the latter from naturalness. What counted as sounding natural, however, was a difficult question.

Sometimes it seems to have meant that music should imitate the sounds of nature, the wind in trees and animal sounds. It is worth recalling the amusing anecdote about the prince who desired to convince his audience that music could be truly natural and thereby of value. He constructed a tent in which he put pigs of all sizes and shapes. To their behinds he fixed pins that were themselves joined to ropes individually attached to the keys of a musical instrument positioned outside the tent. At his party that evening, the prince began to play the instrument, whereupon all kinds of natural squeaking sounds were heard coming from the tent.[26]

Expressing the ideal of naturalism more soberly, Batteux claimed that art has only to use nature's abundant material and therefore needs create nothing of its own. And by nature's materials he could have meant anything ranging from the natural harmonies of the cosmos to constitutive elements of the natural or human world that could be imitated through the medium of word, image, or sound. However understood, the ideal of naturalism was soon to be replaced or at least significantly rearticulated. Theorists began to think especially of the material of music not as natural but as uniquely or specifically

[25] Le Huray and Day (eds.), *Music and Aesthetics*, 237.

[26] Addison often remarks on the use of live animals and animal sounds in London productions of operas. See e.g. *The Spectator*, 6 Mar. 1711. See Neubauer, *The Emancipation of Music*, 70 ff. for further discussion on the imitation of natural sounds.

musical. The arch-formalist Hanslick captured the new ideal, for example, in his renowned statement that the form and content of music together consist in 'tonally moving forms'. It is mistaken, he argued, to think that music is found *in* nature; it is founded only *upon* nature. Musical materials such as melodies may exist in accordance with natural principles, but that does not mean that they themselves are natural.[27]

It became just as familiar, under the influence of the new aesthetic, to talk about musical forms—the sonata, symphony, and concerto— as unique to music itself. Musical form was no longer to be thought of as following the text or the shape of some 'extra-musical' occasion, but as independently designed and independently coherent. As is well known, musicians with both theoretical and practical interests began at this time to give utmost consideration to the development of purely musical forms.[28]

VII

The emphasis on the 'specifically musical' also generated new attitudes towards the tools of art, notably musical instruments. From the earliest times it was held that for music to fulfil its function, it should use instruments unique to its production. Instruments for producing civilized music, be that music edifying, religious, or now aesthetic, were never regarded as profane tools for use in the 'everyday'. They were never arbitrarily used without consideration of their origins, constitution, and effect. Given an interest, for example, in the emotive effects of music, expressed in the Baroque theory of 'affects' (*Affekten-lehre*), instruments were judged according to their ability to represent individual passions. Quantz, who believed that each instrument symbolized a certain human passion, argued that

> to judge an instrumental composition properly, we must have an exact knowledge, not only of the characteristics of each species which may occur in it, but also . . . of the instruments themselves. In itself, a piece may conform both to good taste and to the rules of composition, and hence be well written, but still run counter to the instruments.[29]

[27] Batteux, *Les Beaux Arts*, 31–43; Hanslick, *On the Musically Beautiful*, 68–9.

[28] For a discussion of musical form, see I. D. Bent, *Analysis*, New Grove Handbooks in Music (New York/London, 1987/8); C. Rosen, *The Classical Style: Hadyn, Mozart and Beethoven* (New York/London, 1972); Wiora, *Das Musikalische Kunstwerk*, pt. II, ch 1; and Seidel, *Werk und Werkbegriff*, pt. III, ch. 5.

[29] *On Playing the Flute*, §§ 28, 310.

His view was almost modern.

Under the new aesthetic, however, each and every instrument was increasingly thought of as having its own unique musical colouring. Each played an indispensable part in the performance of the highest form of art, now aptly and proudly, rather than derogatorily, called instrumental music. Each instrument was, for the first time, to be judged on its own musical terms. In a review written in 1809, a critic pointed explicitly to the contribution instruments could make to new specifically musical forms. The symphony, he wrote first, 'has become the highest form of instrumental music'; it has become 'the opera of the instruments'. And then he remarked that, though in the case of the symphony he was listening to (by Friedrich Witt), 'it was a difficult task to explore all the characteristic qualities of each instrument,' still, he was sure that 'each individual instrument was responsible to the whole'. And that, he remarked, was not true of the 'tedious *concerto grosso*' of earlier times.[30]

VIII

The separability principle had been used to formulate a distinction between the musical and extra-musical as music was given its autonomy. It had also been used to redefine the distinction between art and nature. Both uses immediately affected the distinction between civilized and popular music. Once determined by extra-musical criteria, this last distinction was increasingly expressed according to music's degree of separation from the natural world of 'mundane' phenomena, according to the degree of artificiality of its form and matter.

The equation of civilization with artificiality had already been expressed during the Enlightenment. In 1763, theorist John Brown tried to describe how music (still music with words) achieved civilized status by increasing its artificiality.

In the course of time and the progress of polity and arts, a separation of the several parts or branches of music . . . would naturally arise. Till a certain period of civilization, letters and art, the several kinds would of course lie confused, in a sort of undistinguished mass, and be mingled in the same

[30] Schafer, *E. T. A. Hoffmann and Music*, 80. The same critic also remarked that 'everyone knows that instrumental music has now reached heights which could in no way have been conceived even a few years ago'. Note that instrumental music comes to stand either for pure music, or for music produced on instruments as opposed to choral music. Only when understood as pure music is choral music included.

composition, as inclination, enthusiasm, or other incidents might impel. But repeated trial and experiment would naturally produce a more artificial manner, and thus by degrees the several kinds of poem [song] would assume their legitimate form.[31]

The emphasis on artificiality was not always thought to undermine the ideal of naturalness, though in the Enlightenment many theorists thought that such an emphasis did precisely that. Later on, for romantic theorists at least, it was designed to confirm the degree to which the fine arts could be separated from mundane and uncivilized activities, be they activities of making or performance. The term 'artificial', therefore, is best understood in this context less with its original meaning, 'made by human hands', than as 'civilized' or as 'higher', and, particularly in the context of fine art, as belonging only to the domain of fine art and nothing else.[32]

Artificiality so understood was considered by the end of the eighteenth century to be a mark both of the civilized and of the aesthetic. As a mark of the latter, it was crucial to a new understanding of fine art. Let me introduce the idea of *aesthetic remainder* to explain this. It is familiar nowadays to understand the relation between works of art and their physical or natural 'counterparts' as one of 'transfiguration'. The transfiguration of physical or material objects into works of art— what Arthur Danto calls 'the transfiguration of the commonplace' in his book of the same name—appeals to a viewer's interpretative faculties. Interpretation enables a viewer to accept a commonplace object as a work of art, for the process results in the once common-place object acquiring a new set of properties, properties deemed by the institution of art to be aesthetic. These properties constitute the

[31] Le Huray and Day, *Music and Aesthetics*, 85. In 1790, Archibald Alison distinguished airs/songs from compositions by composers on the grounds that songs have a beauty 'discernible by the common people' whereas the latter sort of compositions are of a 'more artificial' nature (*Essays on the Nature and Principles of Taste* (Edinburgh, 1815), i, essay 2, 2. 3. 2: 'Of Composed Sounds, or Music', 272–3.

[32] This present sense of 'artificial' contrasts with a use common in the 18th cent. designating the 'scientific manner of thinking, speaking, and writing' negatively associated with notions of 'enlightenment' and 'civilization'. Many 18th-cent. theorists— Rousseau, Winckelmann, and Schiller—looked back to Antiquity for a sense of the natural and spontaneous to contrast with the artificiality of the modern world. Looking back in this way still captured the idea of transcendence, however, the idea that the best artists and their works could transcend the negative 'artificial' conditions of 18th-cent. society, to meet the Greek aesthetic ideal. Goethe later spoke of Winckelmann as having himself done precisely that, i.e. as being of Greek temperament, as having transcended the conditions of modern man to embrace the values of the past.

aesthetic remainder; strip away (conceptually) the non-aesthetic properties (often an object's use or function) and the aesthetic remains.

The romantic theorists had a formula at hand. The greater the
aesthetic content, the less the worldly content, and, therefore, the
more worthy the art. To increase the value of the fine arts, it was
necessary then to give to them their own forms, materials, means of
production, and so on. The idea was to reduce the number of
commonplace properties. Applying the separability principle, they
managed to make form, matter, and media as aesthetic or artificial as
they could. They found that the procedure was more successful in the
case of instrumental music than in that of any other art. That
instrumental music had the least amount of commonplace or worldly
content, that it fulfilled mundane functions least well, implied that it
had the most pure, aesthetic character. 'It is perhaps in music that the
dignity of art is most eminently apparent,' Goethe wrote, 'for music
has no material element that has to be taken into account. It consists
entirely of form and content; and [therefore] . . . elevates and ennobles
everything that it expresses.'[33]

A truly romantic rationale exists, then, for a transfiguration theory
of art. Transfiguring an everyday object into a work of art depends
(though it need not[34]) on the belief that art has the ability to represent
more or less directly the aesthetic world, a world severed from the
world of everyday objects and concerns, a world more soberly called
'the world of art' or 'the artworld'. Transfiguration also depends upon
a certain kind of illusion, the ability to see or hear in a physical object
or performance, less the concrete or the physical, than the transcendent.
This way of seeing applies not just to twentieth-century, modern art in
which physical objects have been turned into works of art by artists

---

[33] Goethe, *Wilhelm Meisters Wanderjahre* (1829), tr. Le Huray and Day (eds.), *Music
and Aesthetics*, 420. The same argument lies behind Pater's description of the aspiration
of all arts to approximate to the condition of music. 'For,' Pater explains, 'while in all
other kinds of art it is possible to distinguish the matter from the form, and the
understanding can always make this distinction, yet it is a constant effort of art to
obliterate it' ('The School of Giorgione', 86).

[34] Danto provides an explanation that does not explicitly depend upon a romantic
thesis, though it would still be fair to say that many modern views are more sober
expressions of views first put forward in highly romantic terms. For Danto, the core of
the separability principle is the idea of 'aboutness'. Art can say something *about* the
world which the world cannot itself do. Art stands apart from the world in this sense
(*The Transfiguration of the Commonplace*, ch. 3). Cf. Roger Scruton's relevant discussion
of tertiary qualities in his *The Aesthetic Understanding: Essays in the Philosophy of Art and
Culture* (London, 1983), ch. 7, and Wollheim's remarks on seeing three-dimensional
aspects in two-dimensional objects in *Art and its Objects*, § 11.

such as Marcel Duchamp and Carl Andre. It applies as much to traditional works of art in which content and form are developed with inextricable reference to the relevant artistic medium despite their also being embodied in physical material.[35]

IX

To truly aestheticize fine art, to make it completely artificial, the separability principle had to be put to further use. It was employed, for example, in romantic theories of creation and reception. I mentioned earlier the separability principle's involvement in articulating for the creators of art works something approaching a God-like status. When I quoted Körner in support of that view, I indicated that not only artists but receivers of art as well had to separate themselves from the mundane and particular. What did that separation involve?

When Kant developed in the 1790s a systematic theory of aesthetic judgement, it was the principle of separability that motivated him, at least in part. For he intended to give to aesthetic judgement (contemplation and evaluation) an autonomy and uniqueness that would separate it from cognitive and moral reasoning, i.e. from the rules both of theoretical and practical understanding. To achieve that end, he described four moments or desiderata for what he called pure judgements of taste.

(i) Such judgements should be held *universally* for all persons, such that all private points of view are suspended.

(ii) They should be derived from a *disinterested attention*, such that the viewer takes no interest either in the existence of the object or in the particular concept under which the object falls (what kind of object it is).

(iii) Aesthetic judgements should be *purposive yet be without specific purpose*, for beauty is absolute and not instrumental. If an object looks as if it were designed for a moral, practical, or scientific end, and the viewer takes account of that end, then the viewer is not contemplating the object aesthetically.

---

[35] On the level of the aesthetic, and for all works of art, it is the aesthetic content less the ordinary content that is of interest. Note that many artists of the 20th-cent. avant-garde have parodied the romantic transformation by challenging its presuppositions. With so-called 'found art' they have tried, for example, to undermine the distinction between ordinary and aesthetic content, forcing us, in turn, to question the deep-rooted distinction between the ordinary or everyday and the 'higher' levels of the sublime and the beautiful.

(iv) Aesthetic judgements should depend upon a *modality of satisfaction* or sympathy, a reconciliation of all our mental faculties.[36]

These desiderata—especially the first three—helped sever the age-old connection between the arts and their purported external utilities, and that helped guarantee the fine arts their autonomy, though in actual fact Kant did not confine his account of aesthetic judgement to fine art or to any other specified class of objects. Notwithstanding, the general reasoning of the time behind such a severance was this: if the value of art lies in its higher or transcendent content, aesthetic reception must be such that it enables the receiver to grasp this content. Only in the 'pure' contemplation of art is one able to do this. Already in the 1750s, Winckelmann had spoken of his assuming 'a more exalted position' as he contemplated an artwork, so that he could become 'worthy' of the sight. And in such a position, as he described it, he forgot all else.[37]

Forgetting all else meant little more than stripping one's mind of all mundane considerations. During the eighteenth and into the nineteenth century, contemplation was increasingly described, as it was by Kant, as purposeless and disinterested. Such contemplation involved the setting-aside of one's everyday concerns. It was a form of meditation that persons of higher cultivation or taste could successfully engage in. Any judgement of taste, Kant had claimed, 'uninfluenced by charm and emotion, that is based purely on suitability of form, is a pure, aesthetic judgement'.[38] 'The lyre of Orpheus opened the portals of Orcus,' wrote Hoffmann in more passionately romantic terms:

> Music discloses to man an unknown realm, a world that has nothing in common with the external sensual world that surrounds him, a world in which he leaves behind him all definite feelings to surrender himself to an inexpressible longing.[39]

Disinterested contemplation, otherwise described as a free play (*Spiel*) of imagination or fancy, was increasingly described as isolated

---

[36] *Critique of Judgement*, §§ 6–16. Kant further divides aesthetic judgements into those of the beautiful and the sublime. For rather idiosyncratic reasons, he also assigns music a low place in the classification of the arts. Recognizing that music is a 'beautiful play of sensations', he claims that its impression is transitory rather than lasting, and furthermore, that it imposes itself upon the listener for longer than one usually desires. From plastic arts it is possible to turn one's eye away, if one so desires. With music one is often forced to listen whether one wants to or not (§§ 50 and 53).

[37] Winckelmann's reaction to 'Apollo Belvedere', quoted in W. Leppmann, *Winckelmann* (New York, 1970), 154.

[38] *Critique of Judgement*, § 13.   [39] Strunk, *Source Readings*, 775–6.

not only from our everyday concerns, but also from our rational faculties. If aesthetic contemplation was to have any response associated with religious, moral, or emotive 'awe', it was to be elicited through pure aesthetic intuition or through the irrational, and not through some rational or cognitive justification or any other sort of worldly justification. That belief, by itself, constituted a significant break with the past.

X

The separation of aesthetic reception from all other sorts of reception had a direct impact on the autonomous nature of the artwork. Recalling an earlier theme, objects of mundane utility were once called works of art. Under the influence of the new aesthetic, they were called products of craft. The term 'Art' (with a capital A) was to be reserved for products of inspiration and creativity and not used therefore for those of everyday labour. In 1785, Karl Philipp Moritz described the gulf that now separated Art from craft.

> Only when considered in conjuction with their function and as an entity that includes that function do [objects of utility] afford pleasure; divorced from that function, they are a matter of total indifference. . . . The opposite is the case with the beautiful. This has no extrinsic purpose. It is not there to fulfil anything else, but it exists on account of its own perfection. We do not contemplate it to discover what use we may make of it; we use it only to the extent that we can contemplate it.[40]

If the experience of the beautiful was to be severed from the world of everyday concern, the object of contemplation could not contain any feature to threaten this severance. The success of aesthetic reception depended, in other words, upon the work of art's having no referential or external features. Each work had to contain everything of significance within itself.

To explain this idea, theorists began to describe a work of fine art as having its own internal unity. Having such a thing would give the work the kind of self-sufficiency it needed to be an object of aesthetic contemplation. Kant had articulated the point when he spoke of pure aesthetic judgement as being devoid of determinate concepts, and when he spoke of an object's being contemplated not as a particular

---

[40] Letter to Moses Mendlessohn, rpt. Le Huray and Day (eds.), *Music and Aesthetics*, 187.

kind of object existing in the physical world, but rather as an object experienced within, as he said, a flow of representations. Hegel argued, also, that a work of art achieves beauty when it reconciles matter with content, or 'sensuous show' with the 'embodied Idea'. A work's value depends, he said, upon the establishment of an intimacy and union between an 'Idea' and its configuration. From claims such as these, it was a short step to Schopenhauer's well-known view, that works of art are experienced as the objectifications or embodiments of more or less pure aesthetic content.[41]

XI

The belief that works of fine art are self-sufficient, that they bear no external relation to anything else, was finally confirmed as theorists proclaimed that art is an end in itself. Whereas for centuries music, for example, had served as a vehicle for extra-musical belief and sentiment, it now came to be regarded as having its own musical and aesthetic end. Wackenroder concluded in 1799, therefore, that 'the spirit can no longer use [music] as a vehicle, as a means to an end, for it is substance itself and this is why it lives and moves in its own enchanted realm.'[42] But music was not the only art. All the arts came quickly to be seen both as the motivation and as the end of a unique form of human experience—the pure, aesthetic experience.

Danto most recently captured the force of the means–end relation as it applies to the arts when he suggested that if you reduce an artwork to its content, you transcend the medium (the form) through which the content is being expressed. Applied to music the point can be put this way: music traditionally conceived as a vehicle, as functional, vastly underplayed its expressive form. Function made the musical medium essentially transparent by giving priority to that which was being expressed or imitated. It thereby undermined the conception of

[41] Hegel, *Aesthetics*, 39: 'In artistic production the spiritual and sensuous must be as one.' A. Schopenhauer, *The World as Will and Representation*, tr. E. F. J. Payne, i (New York, 1969), § 53, pp. 257 ff. Note that when Kant distinguished pure from impure judgements of taste, he did so by thinking of impure judgements as tainted and influenced by conceptual or other worldly data (*Critique of Judgement*, § 13).

[42] Le Huray and Day (eds.), *Music and Aesthetics*, 250. Cf. Hanslick: 'From an aesthetical point of view, music must be comprehended not so much as cause as effect, not as producer but as product' (*On the Musically Beautiful*, 66).

an object of music as something formed and existing as an end in itself.[43]

Under the new aesthetic, the form and content of works were judged to be merged inextricably: in art it is never simply what you say that counts, but also how you say it. More than this, as embodiments of higher truths, and unlike pragmatic languages, works could not be subject to précis or translation. Everything in a work of art, it was believed, is put there for a reason.

Though an essential move in the emergence of the concept of a work of fine art, the opposition between something's serving an external purpose and being an end in itself could not be taken too literally. The opposition was originally emphasized, as was the opposition between interested and disinterested attention, in order to free art, especially music, from its hitherto functional role. Art, as a non-servile form of expression, was thought to have, in its own right, the capability to express truths either about itself or about the world. This view was captured by Benjamin Constant, among others, in his proclamation of 1804 that art is for art's sake. It was later accurately summarized by Nietzsche when he wrote that 'the struggle against *purpose* in art is always a struggle against the *moralizing* tendency in art, against the subordination of art to morality.' Of course, one could still continue in a more literal vein to regard art in its new emancipated state as having a purpose or function. It was just that, for the first time, its purpose came entirely from within.[44]

XII

The separability principle was used as much in the development of theory as in practice. The principle was employed directly in fact, as

---

[43] *The Transfiguration of the Commonplace*, 151–3. Lessing's *Laocoön* contains an extremely influential discussion of the importance of medium in art. Unlike many of his contemporaries, who sought a single principle for all the fine arts, he stressed the extent to which arts have different possibilities of artistic expression on account of media differences (Nisbet (ed.), *German Aesthetic and Literary Criticism*, 81–6).

[44] On 11 Feb. 1804, Constant writes in his Journal: 'L'art pour l'art, et sans but; tout but dénature l'art. Mais l'art atteint au but qu'il n'a pas' (*Journaux intimes* (n.p., 1952), 58). Nietzsche, *Twilight of the Idols*, tr. R. J. Hollingdale (Harmondsworth, Middx., 1968), §§ 24, 81. For mention of the first uses of this principle, see Beardsley, *Aesthetics*, 284 ff. The principle was also used to indicate art's lack of political responsibility toward and within society. Thus art speaks about art and nothing else. This reading resulted in a complicated tension within romanticism. See A. Comfort, 'Art and Social Responsibility: The Ideology of Romanticism', in R. F. Gleckner and G. E. Enscoe (eds.), *Romanticism: Points of View* (Englewood Cliffs, NJ, 1962), 168–81. I shall return to this issue in Ch. 8.

theorists and practitioners alike began to consider how and where works of fine art should be presented and exhibited. Framing, staging, and placement had to be reassessed, for they were all crucial signs by which an audience would be informed of an object's status, now, as a work of fine art. Signs of this sort came, by 1800, to be associated with public museums, art galleries, and concert halls.

Recently reconceived, museums had come to be regarded in the 1750s as places devoted to object collection and exhibition and not, as before, as mostly private places of learning. The museum was increasingly to give to fine art what Max Weber later called its 'indoor character'. Museum curators would take a work of art and by framing it—either literally or metaphorically—strip it of its local, historical, and worldly origins, even its human origins. In the museum, only its aesthetic properties would metaphorically remain. At the very least, museum visitors would have to be taught to identify the properties they were supposed to attend to in their contemplation of fine art.

This process was thoroughly effective in doing what it was supposed to do: namely, estranging the work of art from its original external function so that its artness would now be found within itself. Or at least, that is what Andre Malraux has concluded. He comments further:

So vital is the part played by the art museum, in our approach to works of art to-day, that we find it difficult to realize that no museums exist, none has ever existed, in lands where the civilization of modern Europe is, or was, unknown; and that, even amongst us, they have existed for barely two hundred years.[45]

The purported autonomy of the fine arts, guaranteed by their placement in museums, raised particularly interesting problems for music. These become apparent as we begin to consider how music came to replicate some of the characteristics of the plastic arts of painting and sculpture. As it entered the world of fine arts, music had to find a plastic or equivalent commodity, a valuable and permanently existing product, that could be treated in the same way as the objects of the already respectable fine arts. Music would have to find an object

---

[45] 'Museum without Walls', *The Voices of Silence*, tr. S. Gilbert (Princeton, 1978), 13. Cf. pp. 53–4: 'The Middle Ages were as unaware of what we mean by the word "art" as were Greece and Egypt, who had no word for it. . . . For this concept to come into being, works of art needed to be isolated from their functions. . . . When art became an end in itself, our whole aesthetic outlook underwent a transformation.' Malraux argues that (for the plastic arts at least, for those are what he is writing about) this process began in the 16th cent.

that could be divorced from everyday contexts, form part of a collection of works of art, and be contemplated purely aesthetically. Neither transitory performances nor incomplete scores would serve this purpose since, apart from anything else, they were worldly or at least transitory and concrete items. So an object was found through projection or hypostatization. The object was called 'the work'.

The projection employed the separability principle, especially when the latter helped found the belief that music could embody higher spiritual truths. With this idea the process of objectification started. Music began to be thought of as partitioned into works each of which embodied and revealed the Infinite or the Beautiful. Each work contained something valuable, something worthy of aesthetic or 'metaphysical' contemplation.

Musical works also began to be marketed in the same way as other works of fine art and, in aesthetic terms, to be valued and contemplated as permanently existing creations of composers/artists. But none of this was straightforward in practical terms. Since music was a temporal and performance art, its works could not be preserved in physical form or placed in a museum like other works of fine art. Music had a problem. It had to replicate the conditions of the plastic arts and, at the same time, render them appropriate to its temporal and ephemeral character. Music resolved the problem by creating for itself a 'metaphorical' museum, an equivalent of the museum for plastic arts— what has come to be known as an imaginary museum of musical works. As late as 1857, Friedrich Theodor Vischer aptly remarked on the fact that 'music's entry into the system' has been so 'prepared by painting that it may be said that her step can be heard at the door. All the essential points in musical theory make it clear that music is on the borderline of the plastic arts.'[46]

---

[46] *Aesthetik oder Wissenschaft des Schönen* (1857) in B. Bujić (ed.), *Music in European Thought, 1851–1912* (Cambridge, 1988), 83. The notion of an 'imaginary museum' stems from Malraux's piece on Le musée imaginaire'—the museum without walls. My usage of this notion does not carry, however, the 'modernist' connotation of stylistic heterodoxy and eclecticism, or of the wide dispersal and availability of art—the 'limitless junkshop'—that has come to characterize the 'modern' museum in this century. (See G. Hough, 'The Modernist Lyric', in M. Bradbury and J. McFarlane (eds.), *Modernism, 1890–1930* (Harmondsworth, Middx., 1976), 316). My usage of 'imaginary' points, rather, to the status of the 'museum' that embodies the normative codes for the post-1800 production of musical works. It refers to the fact that musical works are such that their existence is projected on the basis of the activities of composers, the scores they write, and the transitory performances that result.

The projection of musical works in the imaginary museum required sophisticated thought and strategic action. But the climate of the late eighteenth and early nineteenth century was favourable for the first time, and musicians found that they could at last turn apparent problems to their own advantage. No one better expressed the situation in which music found itself at this time than Kierkegaard:

Music exists only in the moment of its performance, for if one were ever so skillful in reading notes and had ever so lively an imagination, it cannot be denied that it is only in an unreal sense that music exists when it is read. It really exists only in being produced. This might seem to be an imperfection in this art as compared with the others whose productions remain. . . . Yet this is not so. It is rather a proof of the fact that music is higher, a more spiritual art.[47]

Of course Kierkegaard, like Vischer, had the advantage of hindsight. By the time they made their mid-century statements, much of the reconceptualization of musical practice had already taken and fallen into place. What, we need to find out now, had this reconceptualization involved in practical terms? How did the romantic aesthetic affect in practical terms the emergence of the work-concept? What had practice looked like before that emergence and what did it look like afterwards? What bearing did that emergence have on the roles of the composer, performer, listener, and critic, as these were concurrently being redefined in everyday musical practice?

[47] *Either/Or*, tr. D. F. Swenson and L. M. Swenson (Princeton, 1944), 55.

7
Musical Production without the Work-Concept

This chapter and the following one are devoted to contrasting early with modern musical practice—roughly before and after 1800—to show that the work-concept had regulative force in the latter but not in the former. The concept acquired this force as part and parcel of the emancipation of the musical world and the emergence of that world as one concerned with the production of fine art. Historical inquiry reveals that no force for this concept would have been visible or effective had the concept not been delimited, articulated, and specified from every conceivable point of interest. Such points of interest extended well beyond musical and aesthetic theory. They extended far into the depths of musical practice.

This chapter in particular demonstrates how in centuries prior to the nineteenth each moment, aspect, or belief associated with the work-concept was, if not absent, then understood differently. It shows that these moments were not functioning together in such a way as to give the work-concept institutionalized centrality within the practice. It shows, finally, the extent to which the extra-musical understanding of music described in Chapter 5 was expressed and endorsed in some manner on every level of practice.

What, then, did 'early' musicians take to be the object of compositional and performance activities? In what manner was their music received? Where and when was their music performed? Was there any sort of distinction between compositional and performance activities? Who owned music and how was it owned? What functions did notation serve?

I

Despite some appearances to the contrary, musicians did not gain creative freedom until the end of the eighteenth century. Other types of artist had found some independence in the years following 1750, when the Academy in Rome, for example, announced the liberation of artists from guilds. Artists, though not artisans (craft-makers), it was believed, should exercise their imaginations 'freely and with nobility'.[1] Only in 1835 was a Municipal Reforms Act passed in Britain to free musical composers and performers from similar guild restrictions that had been put upon them most notably in the previous century. But had musicians, generally speaking, sought and found any other sort of freedom earlier than this date—perhaps as early as 1750 or even earlier, to match developments in the plastic fine arts?

From earliest times, musicians all over Europe defied the theoretical strictures of their times in minor ways. Forkel noticed ways by which J. S. Bach conceptualized for himself a degree of independence from his day to day service. At the time of his death, Bach was Court Composer to His Majesty the King of Poland and Electoral and Serene Highness of Saxony, Kapellmeister to His Highness the Prince of Anhalt-Cöthen, and Cantor to St Thomas's School. Apparently, though, Bach preferred to designate himself 'Director Chori Musici Lipsiensis' or 'Director Musices', and Forkel believes he did so for a reason. 'Circumstances', he wrote, led Bach 'to emphasize a title which asserted a musical prerogative not confined to the School and the churches it served.' Unfortunately, Forkel did not specify what those circumstances were.[2]

Perhaps a musical prerogative is also found in Bach's career given that he 'did not hesitate to make double use of originally secular works by adapting them to sacred texts, or to include movements from concertos and other instrumental compositions in his cantatas'.[3] Bach recognized the difference in function and importance between sacred and secular music, and between text-based music and instrumental music. He knew that certain kinds of texts and music were appropriate to secular but not to sacred performance. But he also disregarded all such differences, when it came to the seriousness with which he composed *any* sort of music. He considered all acts of composition to be undertaken in 'homage to God'.[4]

---

[1] See P. Gay, *The Enlightenment: An Interpretation, The Science of Freedom* (New York, 1969), 227.

[2] *Johann Sebastian Bach*, 29.

[3] David and Mendel (eds.), *The Bach Reader*, 33. The editors continue: 'The Osanna in the B Minor Mass was originally a movement in a *dramma per musica*, a secular choral cantata performed as a serenade during a visit of the king and queen to the city of Leipzig, but in its later postion within the Mass it forms . . . a glorious piece of religious praise.'

[4] Ibid.

Even so, were such liberties markedly different from those sought by musicians of earlier centuries? I do not believe so. Many musicians before Bach ignored the distinction between sacred and secular music just in the way Bach did. Many adapted music, written first for a secular occasion, for a sacred one. And such a procedure was possible because it was so often the text, less the music itself, that carried the meaning. More importantly, Bach, like composers before him, worked as a musician in the service of church and court, and the terms of employment placed enormous restrictions on his compositional activities. When he used the title 'Director of Music' that meant that he was in charge of musical proceedings, but such charge had its limitations. Two centuries earlier, Monteverdi had had similarly to content himself on being given the same title. Whatever advances in composition and performance were initiated first by Monteverdi and much later by Bach, those composers were still unable for the same sorts of reasons to exercise the independence characteristic of an autonomous and work-based practice.

II

Before the late eighteenth century, 'serious' music was truly a performance art. It was mostly produced in the public arena to perform extra-musical functions. Performances were geared towards the temper and needs of the persons and institutions who determined the functions. Musicians, who were normally in the latter's employ, had little control and power of decision regarding matters of instrumentation, form, length, and text. They obeyed the wishes of their employers. In 1618, Monteverdi was thus obliged to request a favour from his 'Most Illustrious and Most Excellent Lord and Respected Master', whether the latter would be so kind as to let him know 'for how many voices, and how it is to be performed, and whether an instrumental symphony will be heard beforehand and of what kind, . . . [and] on what instruments will it [the canzonetta] be played'. Monteverdi wanted to know all this so that he could, as he put it himself, 'write appropriate music'.[5]

Handel, similarly, did not always know the conditions under which his music would be performed. On one occasion he had so little knowledge of a 'setting in Dublin' that he was forced to adapt the

---

[5] *The Letters of Claudio Monteverdi*, tr. D. Stevens (Cambridge, 1980), 141.

music at the time of actual performance. Even Haydn did not always have enough knowledge of this sort, though of course there were generally accepted procedures for composing occasional music that he, like Handel and other composers before him, was obviously aware of.[6]

The very idea of having to enquire about the conditions for which one is to compose one's music, what sort and how many instruments one is writing for, and how long one's composition should last, is not met with full comprehension today. But once it was quite normal and, furthermore, it made sense. For why should employees of a court or a church be privileged with advance warning or knowledge? The task of musicians was to produce music on request as time and occasion demanded—for the here and now. Bach was thus employed in Leipzig to furnish music for the 'Old Service'. Handel was instructed to write music 'to be disposed after the Manner of the Coronation Service', or to 'perform after the cathedral manner'. C. Ph. E. Bach recalled in the 1770s the 'ridiculous instructions' he was often forced to follow as he composed for 'specific individuals' and 'the public'.[7]

The idea that one first composed a work which then was publicly performed here and there hardly existed. It could not therefore regulate the public activities of composers. This idea hardly even regulated their private activities, when, as C. Ph. E. Bach put it, composers were able to produce compositions in 'complete freedom' and for their 'own use'. For most music privately composed was written for purposes of private exercise, learning, and pleasurable entertainment, so that it was not always even expected that it would travel from one performer's hands to another.[8]

Perhaps it did not matter that musicians had so little control over their public affairs, for they were not always recognized as the authors of their music anyway, and if they were, such recognition was not accorded much importance. Public recognition was given instead to the extra-musical bodies or persons for whom the music was composed; this recognition had status. Thus, in the 1720s, Handel was employed to compose a certain number of operas 'to be written by himself or

---

[6] C. Hogwood, *Handel* (New York, 1984), 168; See Haydn's letter concerning the 'Applausus' Cantata quoted in Sect. 6 below.

[7] David and Mendel (eds.), *The Bach Reader*, 99; Hogwood, *Handel*, 172 and 180; W. S. Newman, 'Emanuel Bach's Autobiography', *Musical Quarterly*, 51 (1965), 371.

[8] 'Among my works especially those for clavier, there are only a few trios, solos, and concertos, that I have composed in complete freedom and for my own use' ('Emanuel Bach's Autobiography', 372).

whoever he chose'.[9] It mattered, but not that much, who composed the music. It mattered much more that the music satisfied or lived up to the demands of the occasion.

Recognition of musical compositions came generally in the form of dedications. Usually, aristocratic or religious dignitaries, or even publishing companies, dedicated the Kapellmeister's music. When musicians did dedicate their own compositions, it was rarely out of a sense of freedom, but out of respect they had for some 'Serene and Gracious Prince' whose employ they enjoyed. John Jacob Heidegger dedicated Handel's opera *Amadigi* to Burlington, in gratitude for 'that Generous Concern Your Lordship has always shown for the promoting of theatrical Musick'. He then added that this opera claimed Burlington's protection in particular because it was 'compos'd in [his] own Family'.[10] That was the point; composing music almost just was a family affair.

The same reasoning gave ownership of music to church, court, or, increasingly, the publishing house, but not to composers. In accepting their subservience, musicians automatically gave up any right to control the distribution of music they composed. They would not even have conceived of composition to be the kind of activity over which they could have property rights. Thus, the contract Haydn accepted in 1761 for employment as Vice-Kapellmeister at the Esterházy court specified that

he was to be under obligation to compose such music as His Serene Highness may command, and neither to communicate such compositions to any other person, nor to allow them to be copied, but . . . shall retain them for the absolute use by his Highness, and not compose for any other person without the knowledge and permission of his Highness.[11]

The sense in which music was owned, however, was variously and often vaguely understood. Typically, it meant that the owner had exclusive rights to publish and distribute music. Alternatively, it could mean that permission was required from the owner to use music of a given occasion (or part thereof) for a different one in another place. Sometimes it amounted to little more than a recognition that some music had been performed at a ceremony or religious service. What ownership did not yet entail was the idea that composers could have completely free and privileged access to the music they composed.

[9] Hogwood, *Handel*, 91. [10] Ibid. 70.
[11] K. Geiringer, *Haydn: A Creative Life in Music* (New York, 1946), 46.

III

The fact that musicians did not own their music, and the fact that music was functional, meant that one musician could make use of any other's music (usually part of it, but sometimes the entire thing) without acquiring permission from the composer, and sometimes even without permission from the owner. Often composers would not even know that someone else had used their music. But, still, this was not considered unfair. Composers could also reuse music they themselves had written for some earlier occasion without always needing the owner's permission. In 1733, when Handel performed his *Water Music*, parts were performed that had been used on other occasions. In 1739, an anonymous writer reported on a public musical performance comprising bits of Handel's operas *Esther* and *Athalia*.[12] Such performances were not those of musical excerpts or medleys of different works, such as one might hear today, as much as they were performances of compositions—sometimes newly named—made up of previously written music. These two examples are not exceptions to the rule. Reusing music in this way was just part of what it meant to compose music.

That music was designed to fit an occasion meant that it had to be adaptable. It had to be adaptable to resident instrumental ensembles, to the occasion, to temporal restraints, and so on. Hence, it was allowed for publishing companies and performing ensembles to alter music at will, as was seen fit. Composers themselves did not hesitate to make these sorts of alterations, either to their own music or to that of others. In 1731, a composition 'formerly composed by Mr. Handel' was revised by him 'with several Additions'. Bach composed a mass formerly composed by Palestrina.[13]

That shared or common use of musical materials was allowed, and even encouraged, was consistent with many contemporary attitudes regarding musical composition. Imitation, for example, understood as copying, involved imitation of another composer's style, as well as the use or recomposition of existing melodies or musical structures. These practices were an accepted way of modelling one's own music upon that of a past master. Corelli adapted for his sonatas themes from Lully's operas; Bach used themes of Vivaldi, Albinoni, Corelli, and Legrenzi. Again, these were not exceptional cases.

[12] Hogwood, *Handel*, 108 and 157.   [13] Ibid. 98.

Alteration of compositions, multiple use of themes, and overlap in thematic structure among various compositions, in addition to the open borrowing of another's music, all effected a free and easy interchange of musical materials among compositions. But that also meant that the musical output of any given composer could be enormous. Vivaldi composed in the space of forty years what we now identify as 845 distinct works; Domenico Scarlatti composed over 550 keyboard works, as well as operas and sacred music; in twelve years, Nicolo Piccinni is said to have written 134 operas, plus oratorios, masses, and cantatas. Haydn's operas were fewer in number: he wrote only twenty-three. But he also composed 106 symphonies and sixteen masses—and more.[14]

In our modern age this number of compositions seems remarkable. Even the low rate of production of Berlioz, who produced only around thirty instrumental works in his lifetime (though he produced other sorts of music as well), seems more normal to us. Nine symphonies is the modern standard—at least for symphonies. There are exceptions of course. Liszt produced about 1,300 pieces of music, though many of these were transcriptions of already existing works, and many of these were composed for easy consumption, in other words, for a popular form of musical entertainment. Charles Gounod composed a similar number. But the general difference in quantity between early and modern practice remains, and makes sense. Any composer's output would be extensive were he employed to compose music, as Bach was, 'on all high holidays and feast days, and any others as they occur, and on the eves of such days, and every Sunday and Saturday afternoon, as well as at the regular Catechism sermons and public weddings . . . in furtherance of divine service to the best of [one's] ability and zeal'.[15] And if some composers were employed to produce music for religious ceremonies, others were hired to do the same for operas, ballets, dinners, soirées, and balls. It is inconceivable to think that any composer would have had time to produce new and original music for each occasion; in practice, they mostly produced music that was appropriate in quantity and type. Thus, in 1719, it is reported

---

[14] 'Bach wrote unceasingly, . . . and the quantity of his works is enormous' (*OED*, 'Work').

[15] Written into the employment contract for services in Halle (David and Mendel (eds.), *The Bach Reader*, 65). Cf. Mozart's comment in a letter to his sister, 4 Aug. 1770: 'In the meantime I have composed four Italian symphonies, to say nothing of arias, of which I must have composed at least five or six, and also a motet' (*The Letters of Mozart and his Family* ed. E. Anderson (3 vols.; London, 1938), i. 226).

that the founders of a musical society at London's Haymarket announced their intention to secure to themselves 'a constant supply of Operas to be composed by Handel'.[16]

To describe musicians as having composed so many individual works is misleading, of course. Many of their compositions would have involved significant overlap and repetition of musical material. And such overlap would not just have existed within a single composer's output, but among compositions by any number of composers. This general sense of borrowing was comparable in fact to the general use of a language for which there is no uniqueness or ownership of any given expression. Musicians could almost say the same thing as often as they wanted.[17]

I say 'almost' because the use of musical material was to a certain extent restricted. Restrictions came in various forms. Musicians would be restricted in the kind of material they could use, insofar as it had to suit the occasion and the text if there was one. Restrictions were also effected when owners of music forbade its further use, when, say, some music had come to take on a particular religious significance. Recall Mozart's retrieving of the score of Allegri's *Miserere* contrary to papal decree. The other type of restriction was articulated through a rough and ready sense of what it meant for someone to copy, steal, or plagiarize music written by another composer.

The free and easy interchange of musical expressions, composed to suit any and all occasions, had entailed its appropriate terms of ownership or the lack thereof. But the fact that there was this sort of interchange never meant that composers could make use of already existing music claiming as they did so that they composed the music from scratch. In 1731, following an 'acrimonious' debate in the Academy of Ancient Music, a composer named Bononcini was discredited for doing just this, for claiming authorship of a madrigal, 'In una siepe ombrosa'. The madrigal was found to have already been printed in a collection by Lotti.[18]

---

[16] Hogwood, *Handel*, 76.

[17] Leonard Meyer recently described the difference between early and modern output this way: '[I]f Bartók and Schoenberg composed fewer works than Mozart, it was not necessarily because they were less gifted than he, but because the styles they employed required them to make many more conscious, time-consuming decisions. Mozart could compose with astonishing facility partly because the set of constraints he inherited (and which he partly modified), the so-called Classic style, was especially coherent, stable, and well-established' (*Style and Music: Theory, History, and Ideology* (Philadelphia, 1989), 5).   [18] Hogwood, *Handel*, 96.

But consider, for the sake of contrast, the story of one James Kent (1700–76), about whom it was reported after 1800 that he had not only adopted 'the style of Dr. Croft, but [he had] even borrowed his ideas, and avowed it, as if it were a matter of course'. It was a matter of course in Kent's time, though the ninteeenth-century reporter George Hogarth fails to recognize this. The story continues. Kent 'once said to a friend who was present at the rehearsal of one of his anthems: "I know your thoughts: there is the same passage in Dr. Croft; but could I have done better than copy him in that case?" '[19] Presumably Kent thought a great deal of the passage.

As early as 1722, Mattheson reckoned it important to mention that Handel had used some of his music.

In the *opera Porsenna* of my composition, as it was performed here 20 years ago [actually 1704], and *accompanied* by *Haendel* under my *Direction*, is found an *Aria* whose opening words run: *Diese Wangen will ich kussen*. . . . It can well be that the melody may have seemed not unacceptable to *Haendel*: for not only in his *Agrippina*, which appeared in Italy, but also in another, new *opera*, recently performed in England and treating of *Mutio Scaevolo*, he has chosen just this same melody, almost note for note.[20]

Unlike Bononcini's case, Mattheson did not criticize Handel outright for having borrowed the melody in question. Handel did just what James Kent did later on. He used a passage he liked that was composed by someone else, and he used it, apparently, on numerous different occasions.

In 1719 Jonathan Richardson provided a defence of common use of material in the arts, the force of which (if not Richardson's actual text) probably influenced Mattheson's understanding of these matters. He wrote:

Nor need any man be asham'd to be sometimes a Plagiary, tis what the greatest Painters, and Poets have allowed themselves . . . indeed 'tis hard that a Man having had a good Thought should have a Patent for it for Ever. The Painter that can take a Hint, or insert a Figure, or Groupes of Figures from another Man, and mix these with his Own, so as to make a good Composition, will thereby establish such a Reputation to himself, as to be above fearing to suffer.[21]

[19] Hogarth, *Music History*, 117.
[20] Reported by Hogwood, *Handel*, 46.
[21] *A Discourse on the Dignity, Certainty, Pleasure and Advantage of the Science of the Connoisseur: Two Discourses*, ii: 'An Argument on Behalf of the Science of the Connoisseur' (London, 1719), quoted in Hogwood, *Handel*, 275.

Though he makes few references to music in his text, Richardson's words portray the attitude towards the shared employment of musical language that prevailed in centuries prior to the nineteenth. And when Handel's music was much later criticized, mostly after 1800, on the grounds that it too often made use of other composers' materials, many critics none the less forgave him on Richardsonian grounds. Handel was above 'fearing to suffer' for, though he borrowed the music of others, he usually took, the critics said, 'other men's pebbles and [polished] them into diamonds'.[22]

Apparently not everyone was willing to forgive Handel. Expressing the rife competition that had arisen between supporters of Bach and those of Handel at the beginning of the nineteenth century, Samuel Wesley wrote precisely to the point. Criticizing Handel for making his reputation in England 'wholly Constituted *upon the Spoils of the Continent*', he commented on the fact that

[t]his would nettle the Handelians devilishly; however it is the strict truth, for we all know he [Handel] has pilfered from all Manner of Authors . . . and although it is certain that what he has taken he has generally improved on (not when he robbed the Golden Treasury of Sebastian by the way), yet there is such a meanness in putting even his own subjects in so many different Works over and over again.

Handel, Wesley concluded, 'has as little just claim to the merit of original genius as the most servile of his Imitators'.[23] Memory of what the conditions of practice had so recently been like was clearly very short.

IV

The idea of a work of music existing as a fixed creation independently of its many possible performances had no regulative force in a practice

[22] Hogwood, *Handel*, 266. As part of the development of national musical styles, the 18th cent. witnessed much debate about whether and with what effect German composers should borrow French and Italian styles, the French borrow German styles, etc. Much of the criticism was marshalled against the borrowing by one composer of another's music. Yet the criticism was not directed against borrowing per se, but against borrowing music of a foreign style. Thus in 1746, an article entitled 'La Corruption du goût dans la musique français' mentioned the invasion of musical styles developed by Telemann, Vivaldi, and Handel as 'baleful foreign corruptions' (W. Mellors, *François Couperin and the French Classical Tradition* (New York, 1968), 280).

[23] *The Wesley Bach Letters: A Facsimile Reprint of the First Printed Edition* ed. P. Williams (London, 1988), 9–10. Wesley also complains that C. Ph. E Bach had robbed his own father (22).

that demanded adaptable and functional music, and which allowed an open interchange of musical material. Musicians did not see works as much as they saw individual performances themselves to be the direct outcome of their compositional activity. Mozart composed his 'Linz' Symphony , K425 when he realized that no music in his possession was suitable for a forthcoming occasion. As he wrote to his father on 31 October 1783: 'I am giving a concert in the theatre here and, as I have not a single symphony with me, I am writing a new one at break-neck speed, which must be finished by that time.'[24]

Music was not always produced to outlast its performance or survive more than a few performances. And when it did survive many performances, numerous changes could and usually would be made to the music in the process. Rarely did musicians think of their music as surviving past their lifetime in the form of completed and fixed works. When musicians thought about repeatability, they thought more of the multiple uses of themes and parts for various different occasions, than of one and the very same whole composition being repeated in performances dedicated to the performing of that very composition.

Since music was mostly appreciated by the public because it served this or that occasion, no one was much concerned about its longevity. Thus, write David and Mendel,

[t]he composer of Bach's time was hardly more concerned with posthumous fame than the medieval artist had been. The cantatas and Passions Bach performed year in and year out in the Thomas-Kirche . . . were mostly of his own composition. Bach had good reason to assume that his successors would not perform the works he wrote any more frequently than he performed compositions written by his predecessors; and indeed his works were as promptly laid aside when he died as theirs had been. . . . [T]here is no indication either in his own writings or in contemporary comments that his works were considered timeless or even likely to survive their author. . . . In this respect, as in so many others, Bach was a thoroughly unromantic figure.

Furthermore, they add, no composition by Bach was deemed worthy of re-publication in full form between 1751 and the early nineteenth century. At most, parts of his compositions were published and distributed.[25]

---

[24] *Mozart's Letters*, ed. E. Blom (Harmondsworth, Middx., 1956), 215.
[25] David and Mendal (eds.), *The Bach Reader*, 43 and 359. Until the 'Bach Revival' at the end of the 18th cent., Bach was remembered as having been a good organist and contrapuntist. See Wesley, *The Wesley Bach Letters*, p. xvii.

Even in 1791, following the annual Handel Commemoration in London, Burney remarked that

[i]t cannot be reasonably supposed, that there is a sufficient number of persons in this kingdom, whose wealth, or zeal for the honour of this great and favourite master [Handel], will continue much longer to enable or incline them to attend such an expensive performance year after year merely to hear the same pieces repeated.[26]

Obviously his values were not yet modern. But he does none the less appear to appreciate novelty. During his lifetime he had witnessed more musical events being given not, in Handel's terms, 'as an act of religion' or as an act of ceremony, but for the purposes of a pleasurable 'Publick Entertainment'. Novelty, not to be equated with what became in the romantic aesthetic a concept of artistic originality, had increasingly become a central demand, either of those who favoured a more popular form of 'classical' music entertainment, or of those for whom a repertoire of musical classics worthy of repetition was not yet an entrenched idea. However, even if, *de facto*, novelty was a feature of much early music production (which of course it was—even Plato recognized that), like early counterpart ideals of innovation and originality, it was supposed to have negligible regulative force in the production of 'serious' music.

V

The assumption as to the occasional and transient nature of perform-ances undermined the need for what we now take to be a fully specifying notation. In early centuries, say, in the sixteenth, it had been believed sufficient to notate the figured bass and the melodic outline, leaving the performance to embellish and perform extempore according to established conventions and taste. Even in the eighteenth century, performers used well-established and traditional conventions for reading incomplete scores. Despite the latter's increasing com-plexity, performers were still required to fill in requisite embellishments according to contemporary principles of taste.

Performers did not generally play music, then, with the idea of instantiating an already completed work, completed in its every structural component. Like composers, they performed with the understanding that they had an extra-musical function to fulfil. They

---

[26] Reported by Hogwood in *Handel*, 245.

treated music pragmatically, as a language or medium for use (an attitude which had no bearing, incidentally, on the beauty of the music produced). According to a more or less detailed outline, with more or less detailed instructions for performance, provided by the composer, the performer filled in and completed whatever musical expression was required. Gerald Abraham provides the apposite description. 'Not only during the Middle Ages,' he writes, 'but for centuries after them, the conception of a single "correct" method of performance [of a given composition] did not yet exist.' There were obviously 'inadmissible ways' of performing, Abraham continues, but there was a choice among 'admissible ones'.[27]

Considerable tension was the immediate consequence of unstandardized and incomplete notation, and it was apparent early on. In 1716, Couperin complained of the fact that

we write otherwise than we perform, whereas the Italians on the contrary write their music in the true values which they have conceived for it. For example, we dot [in performance] several quavers [which follow] in succession by step: and yet we notate them equal.[28]

A few years later, Roger North remarked that

it is the hardest task that can be, to Pen the Manner of artificiall Gracing an upper part. It has bin attempted, and in print, but with woefull effect. . . . The spirit of that art is Incommunicable by wrighting, therefore it is almost Inexcusable to attempt it.[29]

Composers in the eighteenth century were already beginning to demand that performers play in such a way as to comply with their scores. But as long as the composers provided incomplete or inaccurate scores, the idea of performance extempore could not acquire its distinct opposite, namely, the fully compliant performance of a work. Such a contrast emerged fully around 1800, just at the point when notation became sufficiently well specified to enable a rigid distinction to be drawn between composing through performance and composing prior to performance.

Bach was appreciated as much for composing through performance, i.e. extemporizing, as he was for any other form. Mizler reported in 1738:

[27] *The Concise Oxford History of Music*, 105–6.
[28] Donnington, *Baroque Music*, 44.
[29] 'The Art of Gracing', in *Roger North on Music*, 149.

Whoever wishes truly to observe what delicacy in thorough bass and very good accompanying mean need only take the trouble to hear our Capellmeister Bach here, who accompanies every thorough bass to a solo so that one thinks it is a piece of concerted music and as if the melody he plays [in the right hand] were written out beforehand. I can give a long testimony of this since I have heard it myself.[30]

When Mozart and Clementi engaged in an extemporization competition in 1781, few present, if any at all, thought that extemporization was a strict alternative to the pre-composition of music, and few degraded it for being of lesser value than other forms of composition. Indeed, it was considered an enormous achievement and convenience to be able to produce well-composed music on comand. Roger North records the praise that a certain Mr. Prencourt (Master of the Chapel to King James II's court) received as 'the best and quickest composer that ever they knew'.[31] Such a resource, incidentally, also fulfilled many a person's desire for novelty.

Comparable was the respect accorded to eighteenth-century composer-performer virtuosi, who were able to demonstrate their talents most immediately in extempore performance. Unlike practices stressing conformity to pre-established rules, the eighteenth century increasingly accepted virtuoso and extemporized performance as a form of public entertainment. In this entertainment, musicians could show their skills in performance as well as in composition. But the point is that 'virtuoso' was still a term used as much in regard to compositions performed as premeditated performance, as to performance extempore.

VI

That composers did not provide full or through-composed scores for their compositions meant that they had limited control of their performances, and that left them with two options. Either they had to accept that they be present at every performance to ensure that their music was played in something like a satisfactory manner and according to their wishes. Or they had to accept that, in their absence, they might have to send instructions to performers and be content with the hope that their letter would be read and acted upon. In 1768, Haydn was obliged to send a letter of instruction to an Austrian Monastery regarding its performance of his 'Applausus' Cantata. 'Since I cannot

---

[30] David and Mendel (eds.), *The Bach Reader*, 231.
[31] *Roger North on Music*, 52.

be present myself at this ["Applausus"],' he began, 'I have found it necessary to provide one or two explanations concerning its execution.' What followed were ten very detailed instructions for the performance. His eleventh comment was part request, part apology.

I ask everyone, and especially the musicians, for the sake of my reputation as well as their own, to be as diligent as possible: if I have perhaps not guessed the taste of these gentlemen, I am not to be blamed for it, for I know neither the persons nor the place, and the fact that they were concealed from me really made my work very difficult.[32]

Had composers provided full scores of their compositions, they would have given themselves another option, the option to detach themselves from performances altogether. But they did not do this. So, for much of the eighteenth century, composers and performers alike continued to see the difference in their roles to be more marginal than we do today, and to be in significant part a matter of control. Composers usually took charge of musical ensembles in their role as performer-leaders or conductors. Like performers, however, they were restricted by the demand to perform music at the right place for the right amount of time. It was not as if composers produced works and performers produced performances.

The continuity rather than distinction between compositional and performance activities is further revealed in certain expressions with common currency in the eighteenth century. When Handel speaks of composing performances 'of musical entertainments' and of persons 'performing in Musick', one gets a sense of just how continuous the activities were.[33] In a comment made in 1739, look at how the author here slides from speaking of the performance to speaking of the composition itself.

I have never yet met with any Musical Performance, in which the Words and Sentiments were so thoroughly studied, and so clearly understood; and as the Words are taken from the Bible, they are perhaps some of the most sublime parts in it. I was indeed concern'd that so excellent a Work of so great a genius was neglected, for tho' it was a Polite and attentive Audience, it was not large enough to encourage any future Attempt.[34]

---

[32] *The Collected Correspondence and London Notebooks of Joseph Haydn*, ed. H. C. Robbins Landon (London, 1959), 9–11.

[33] Hogwood, *Handel*, 108. Cf. *Roger North on Music*, 24.

[34] Hogwood, *Handel*, 157.

Comments about performances and those about works are often confused today. But the point is that to speak in this way is regarded as a confusion. Now we are taught to take care not to confuse judgements made about performances of works with those made about the works themselves. Before 1800, however, this critical difference was not expressed, or if it was, it was not much worried about.

When the Earl of Egmont recalled in 1741 that he 'went to Lincolns Inn playhouse to hear Handel's music for the last time, he [Handel] intending to go to a Spa in Germany', he expressed his belief both that music did not survive the presence of its composer, and that composers were usually present and involved at the performance of their own music. Both were common beliefs. To hear a composer's compositions was usually to hear the composer perform them. Even in 1802 Prince Nicolaus II of Esterházy still had to agree with Haydn that 'it would be very difficult—especially in the case of new works—to perform music without the personal direction of the composer'.[35]

How against the norm Handel and Bach would have thought it that musicians could produce music the performance of which they were not at all involved in, or might have cared little about, or that perhaps was unplayable. All these possibilities, each of which makes sense because we sharply distinguish the activity of creating permanent works from the production of their transitory performances, were to be articulated and realized later on.

VII

In the eighteenth century musical performances often took place with several interruptions. On occasions they were 'interrupted' by non-musical parts of the religious service, though it is rather misleading to call these interruptions. Real interruptions did occur, however, in secular entertainments and sometimes even in religious events, and for several reasons. Often performers made enough mistakes for the performance to have to be repeated. 'False starts' were a common occurrence. Often audiences became bored by the music, so that it did not always get played through to the end. Often there were long breaks or intervals between the playing of pieces or of parts of a given composition. Often single parts (an air, or a duet, or a saraband) of a given composition would be performed on their own. The overall

[35] Ibid. 166; Haydn, *The Collected Correspondence*, 206.

result was that musical compositions were rarely played from start to finish in a single performance.

Compositions were interrupted partly because they were not performed in concert halls devoted to their performance. Usually musical performances were background affairs within a church or court. As accompaniment either to serious or frivolous activities, they were rarely the immediate focus of attention. That fact was obvious given the behaviour of their audiences. Even the term 'audience' is misleading here, for music was not so much listened or attended to, as it was worshipped, danced, and conversed to. It was quite to be expected that audiences would applaud, chatter during, and sing along with a performance.

Thus at one of Handel's performances there were 'shouts and acclamations at almost every pause'. At another time there were several 'disorders interrupting' the performance.[36] Burney recalled the time when

Signor Pugnani played a concerto this morning at the King's chapel, which was crowded on the occasion. . . . [But] he did not appear to exert himself; and it is not to be wondered at, as neither his Sardinian majesty, nor any one of the numerous royal family seem at present, to pay much attention to the music.[37]

Even in more 'civilized' circles it was a rare event for an audience to remain silent for an entire performance or a substantial part thereof, causing Mozart, for example, to complain on many an occasion that no one listened to his music. He had reason to complain. Few persons did listen attentively to his or anyone else's music.

Performances were interrupted by performers and audiences, if not just by the style and shape of the occasions themselves, then because the extra-musical generally had priority over the musical. Musicians were not in a position to demand that it be otherwise. But then they did not generally expect for most of the eighteenth century, and few others did either, that one would hear a public performance in order to hear a pre-composed, completed work which was performed just for the aesthetic sake of performing and hearing that work. The rationale and success of a performance rested predominantly on whether an audience had been efficaciously affected in a manner appropriate to the occasion.

Even the idea of an ensemble's *rehearsing* a composition for adequate performance did not meet with much enthusiasm or comprehension.

[36] See Hogwood, *Handel*, 45, 83, and 158.   [37] *Musical Tour*, 59.

The term 'rehearsal', and its equivalents '*die Probe*', '*die Wiederholung*', and '*die Aufzählung*', were often used interchangeably with terms designating performance. One would hear of a public or private rehearsal of some music, where this was not normally, as it normally is now, a practice session, but the actual performance of the music.[38] Increasingly, during the eighteenth century, rehearsals came to mean preliminary 'play throughs'. Later they came to be conceived as private occasions for concentrated practice and learning. It was not until very late in the century, however, that the concept of a rehearsal as practice came to be sharply distinguished from a performance, and, furthermore, to be deemed necessary for adequate performance.

In 1784, Burney suggested that the fact that a rehearsal was taken before a performance of Handel's music was 'indisputable proof of the high state of cultivation to which practical Music is at present arrived in this country'.[39] But it was not yet the norm. In 1795, Haydn recorded in his *London Notebooks* that he had been 'invited by Dr. Arnold and his associates to a grand concert in Free Masons Hall'. He continued: 'One of my big symphonies was to have been given under my direction, but since they wouldn't have any rehearsal, I refused to cooperate and did not appear'.[40] In 1811, Thomas Busby still felt the need to assert in his *Dictionary of Music* that 'rehearsals, especially of new music, are indispensably necessary'.[41] For reasons having to do with interpretative fidelity and perfect compliance, rehearsals would soon be deemed necessary for the performance of works, old as well as new.

VIII

That performances were generally rather haphazard and unrehearsed points to the extent to which music was heard not for itself but in service to something else. The same conclusion is demonstrated by the fact that if there were written programme-notes for performances at all, and if such notes were distributed to audiences as well as performers, they often indicated the order of activities—of both a

---

[38] Cf. Hogwood, *Handel*, 118 and 156.　　　　[39] Quoted ibid. 238.

[40] 30 Mar. 1795, Haydn, *The Collected Correspondence*, 289.

[41] *OED*, 'Rehearsal'. Many accounts of insufficient rehearsal time are given in Mozart's letters: e.g. the letter to his father of 3 July 1778: 'You have no idea how they twice scraped and scrambled through [the Symphony K297]. I was really in a terrible way and would gladly have had it rehearsed again, but as there was so much else to rehearse, there was no time left' (*The Letters of Mozart*, ii. 825).

musical and non-musical sort—rather than, as later would be the case, the order of musical works to be performed.

Thus, for example, they sometimes indicated the order of a 'Divine Service', or of an evening's entertainment, with no sense that the programme should be confined just to musical items. Items were also sometimes referred to in performative rather than substantive terms. For a service in 1714, notes written by Bach (albeit private ones) indicate that he would begin by 'preluding', after which a 'motetta' would be sung, and then after much 'intoning', 'singing', 'preluding', and 'reading', he would 'prelud[e] on the composition'—probably on the Cantata—after which there would be 'alternate preluding and singing of chorales until the end of Communion'.[42] It was quite common also to read of a given person singing an aria, which would then be followed by another person playing, say, a minuet on the piano.

IX

Rehearsals were uncommon partly because professional orchestras hardly yet existed. They began to exist in the last thirty years of the eighteenth century. That musical ensembles were generally in the employ of a church or court often entailed, as a matter of fact though not of necessity, that standards were very low. Hence the innumerable 'false starts'.

For most of the eighteenth century, composers were not only performers but conductors as well. Conductors were thus more commonly known as *Kapellmeister* or *chefs d'orchestre*. The activity of conducting was conceived primarily in terms of straightforward time-beating and not, as it was to become later, as predominantly an aesthetic matter of expression and interpretation. 'The keeping of the beat', Mattheson wrote in 1719, 'is the principal function of the director.' The director was to acquire the knowledge necessary to 'direct a secular or sacred musical organization.' His responsibility was,

---

[42] David and Mendel (eds.), *The Bach Reader*, 70. Cf. James Grassineau's *A Musical Dictionary: A Facsimile of the 1740 London Edition* (New York, 1966), where 'Musica Poetica' is defined performatively, as 'the art of inventing songs, or modulating concords or discords together agreeably, and makes what we call composition'; 'Composition' is defined as 'the art of disposing musical sounds into airs, songs, etc., either in one or more parts, to be sung by a voice, or played on instruments'; 'Concerto' or 'concert' is defined as 'popularly a consort, a number or company of musicians playing or singing the same piece of music or song at the same time'.

as Rousseau put it, 'régler le mouvement et marquer la mesure'. He was expected to guide an ensemble with the idea of ensuring 'a finished performance', in much the same way as a controller with a baton directs the traffic.[43] In an era when performances were unrehearsed, getting to the end without too many breaks, mistakes, or interruptions was just about the Kapellmeister's most serious concern.

At this time it was not yet the norm, however, that a time-beater should even be present at a performance, confirming an idea once put forward by Hans Keller that early ensembles managed, as it were, on their own.[44] Often principal violinists or harpsichord players determined and kept the beat. Often the performers were also the composers of the music. Apparently these ensembles managed quite well. 'The opera in Paris', Rousseau recalled, 'is the only theatre in Europe where they beat time without keeping it; in all other places they keep time without beating it.'[45]

If conductors were present at a performance at all, the manner in which they gave the beat was completely unstandardized. Despite existing treatises discussing the pros and cons of different time-beating patterns, the way conductors waved their arms, and the particular number of beats indicated for any given bar, remained more often than not a matter of personal taste. Usually, the 'audible' conductor banged or waved the beat with a stick, a piece of paper, or a handkerchief, on the floor, the music stand, or in the air. 'How much our ears have been shocked at the Opera in Paris,' Rousseau wrote in 1768, 'by the continual and disagreeable noise made by the person who beats time with his stick.'[46] The audience, however, did not suffer nearly as much as certain conductors who managed to injure themselves with their sticks, or, in the case of one conductor, to cause his own demise. All such conditions were either to change or to alter in significance after 1800.

---

[43] Galkin, *History of Orchestral Conducting*, 187, xxxiv, and 195. Galkin not only provides a thorough history of conducting, from which I have derived many of my own remarks on this activity, but he does so deliberately to characterize the differences between pre- and post-1800 conducting.

[44] *Criticism*, ed. J. Hogg (London, 1987), 21–5. With this remark, Keller intended to show the 'phoniness' of the modern conducting profession. I have discussed his view in detail in my 'The Power of the Podium', *Yale Review*, 79: 3 (1990), 365–81.

[45] Galkin, *History of Orchestral Conducting*, 443.

[46] Ibid. 443 and 191. Cf. *Roger North on Music*, 106: in consort, 'nothing like a roll of paper in the hand of the artist' is better, 'without noise and above board'.

X

When Haydn expressed in 1790 the difficulties he was experiencing composing a sonata for Maria Anne von Genzinger, because, as he wrote, he was 'no longer accustomed' to the 'capabilities of her piano', he pointed to the extent to which musicians of his time still composed music for actually existing instruments and specific performers.[47]

It was not usually possible for early musicians to function with any sort of distance from the everyday demands that their practice put on them. This is not meant to imply that they had no sense of a perfect sound (*Klangideal*) for their compositions; they did. Neither is it meant to suggest that musicians did not get frustrated with bad instrumental players and low-quality and inadequate numbers of instruments. Bach got into quite a brawl after calling a performer a 'nanny-goat bassoonist'.[48] The point is just that composers did not usually have the advantage of conceiving of their music at a distance from the particularities of its immediate performance. The only sort of distance they did have generated an obligation on their part to specify instruments in very general terms, in terms of pitch ranges. This was a direct consequence of their having to produce music immediately adaptable to any conditions under which they or others might be forced to perform. After all, musical ensembles were not yet standardized either in size or kind.

When Couperin did not specify the instrumentation or medium for performance of his *Concerts royaux* (suites composed 'to soften and sweeten the King's melancholy'), it was because almost everyone involved in their performance would already have known which instruments were suitable. Not only were pitch indications sufficient to determine the suitability of instruments, but instruments themselves were also still being identified and judged suitable according to the emotions their playing could trigger.[49]

The kind of distance which early musicians did not have, and which modern composers gradually acquired, was one enabling them first to determine the kind of instrument they wanted their music to be played on and then to expect that their determination would be

---

[47] Haydn, *The Collected Correspondence*, 106.

[48] David and Mendel (eds.), *The Bach Reader*, 51.

[49] Mellors, *François Couperin*, 234. In 1730 Peter Prelleur remarked that 'Organo signifies properly in Organ, but when it is written over any Piece of Musick, then it signifies the Thorough Bass [leaving undetermined the actual choice of instruments].' The bracketed comment is added by Donnington (*Baroque Music*, 155).

followed. Early musicians did not have the privilege of expecting performances to fit their music. They generally had to content themselves with fitting their music to performance. Thus, Mozart mentions an order he received to have an opera he had composed transcribed for performance in Lisbon. He does not seem to have had any choice in the matter.[50]

When treatises on instruments were written respectively for flute, clavier, and violin by J. J. Quantz, C. Ph. E. Bach, and Leopold Mozart, their content comprised discussions of technique and musical principles for adequate performance of music on particular instruments. Their concerns were, literally for all intents and purposes, practical. It would be misleading, therefore, to claim that Quantz wrote a treatise on instrumentation as such with a focus on the flute. Rather, as the title *Versuch einer Anweisung die Flöte traversiere zu spielen* indicates, he wrote a treatise on flute-playing.[51] Not much had changed in this regard since Couperin wrote his *L'Art de toucher le clavecin* in 1719. Many treatises were written in the same period to defend certain instruments, which were supposedly being surpassed, in physical and aesthetic quality, by newly invented ones. Hubert le Blanc's *Défense de la basse viole* was written just to defend an instrument which he believed was still capable of conveying the subtlest emotional nuance.

Only much later were composers able to consider matters pertaining to instrumentation freely and without restraint at a comfortable distance from particular performers, from particular instruments, and from functional concerns, such that they could consider the contribution of instrumental colouring to the aesthetic character of their works.

XI

Publication procedures of the eighteenth century reflected every aspect of a predominantly functional and pre-work-based conception of music. Published music referred on title-pages not just to composers but also (and with greater emphasis) to the occasions for which the music had been written, the person or persons for whom it had been written, and the date of its performance. On the title-page of Bach's first published composition we are informed that the composition is a

[50] 12 Jan. 1771, *The Letters of Mozart*, i. 265.
[51] The same applies to Mozart's *Versuch einer gründlichen Violinschule* (Augsburg/ Wien, 1756) and C. Ph. E. Bach's *Versuch über die wahre Art das Klavier zu spielen* (Berlin, 1762).

Gratulatory Church Motetto as given when at the Solemn Divine Service in the Principle Church B.M.V. with God's blessing the Council was Changed on the 4th of February in the year MDCCVIII and the Government of the Imperial Great City of Mülhausen was joyously entrusted to the Fatherly Care of the New Council namely to the Most Noble, Steadfast, Most Learned and Most Wise Gentleman MR. ADOLFF STRECKER and the Noble, Steadfast and Most Wise Gentleman MR. GEORG ADAM STEINBACH both Most Deserving Burgomasters as well as the Most Highly Respected Members most dutifully furnished by Johann . . .

Publication procedures also revealed the lack of a rigid distinction between work and performance. Manuscripts, for example, were dated in such a way as not to distinguish between the date of compositional activity and that of performance. Nowadays, we date a work according to the time (year) of completion of composition. Obviously many works are completed just before their first performances take place, in which case the dating of the two is coincident. At one time, by contrast, compositions for public occasions were dated according to the time of performance. This made sense. If one composes music for a specific performance, and one sees the performance as the completion of one's activity, one would expect the time and date of composition to coincide with that of the performance in principle and not just in practice.

Thus Haydn, for example, dated his Symphony No. 9 as 1795 to mark the time of its first performance, even though he actually 'completed' the composition of the symphony two years earlier in 1793.[53] Other composers, earlier or contemporaneously, could have done the same thing, of course, and perhaps their action would have found the following explanation. Even if musicians were beginning to see composition as an activity that took place quite independently of actual performance activity, they might still have continued to see the former as truly completed only in the latter. (I say 'actual' so as not to exclude the fact that part of what it now means to create a work is to prepare the work for its performance.) This attitude would not even have interfered with the common practice whereby in the event that music was prepared for a performance that for some reason was delayed or cancelled, the composer would use the music temporarily elsewhere. In modern times, by contrast, were one to compose a work that was performed for the first time two years after the compositional

---

[52] David and Mendel (eds.), *The Bach Reader*, 57.
[53] Haydn, *The Collected Correspondence*, 146.

activity had ceased, one would probably think of the work as completed yet unperformed. And one would probably leave the composition alone until it did finally receive the performance 'owed' to it.[54]

In contrast to manuscripts produced for public performances, those for private performance could not always be dated at the time of first performance, especially if they were not composed for a specific and single public performance. An interesting alternative procedure was offered by Bach, when he dated a manuscript to indicate the time when he began to compose the music, rather than the time when he felt the activity to be completed. On a title-page he wrote, 'Clavier-Bürslein vor Wilhelm Friedemann Bach, angefangen in Cöthen, Jena 22. Januar, Anno 1720.'[55]

Published music not only recorded the fact that music had been performed on a given occasion, if it had been, but also that most music performed in public places was choral rather than instrumental. Composers themselves were often publicly referred to as musical orators, musical poets, and dramatic composers, and their compositions as musical dramas. To be sure, instrumental music was published as well, but the interest in doing so was not what we would expect it to be. Publishing such music did not mean publishing individual compositions of instrumental music. It meant publishing sets, collections, or series of such.[56]

The explanation for this was that instrumental compositions were rarely publicly performed as separate entities in their own right. They were rarely performed independently, but as they had been for many centuries, as preludes, interludes, postludes in the service of larger choral, dramatic, or operatic compositions. Early symphonies were typically instrumental overtures to these larger compositions and were usually referred to as such—almost until the end of the eighteenth century in fact. Rousseau provided the canonical definition of an 'overture' as a

symphonic piece which one attempts to make splendid, imposing, harmonious, and which serves as an introduction to operas and other lyric dramas of a certain length.[57]

---

[54] The difference in conception is sharply drawn here for the sake of argument. Even modern composers would feel some sense of incompletion and regret were their works to be performed for the first time only years after they had composed them.

[55] David and Mendel (eds.), *The Bach Reader*, 79.

[56] Cf. David and Mendel: 'It was the custom of his time to create and publish series of works, not individual compositions' (*The Bach Reader*, 41).

[57] 'Pièce de symphonie qu'on s'efforce de rendre éclatante, imposante, harmonieuse, et

Late in the eighteenth century, influenced by the compositions of Stamitz, Haydn, and Mozart, the terms 'symphony', 'sonata', and 'concerto' gradually began to be used to refer to independent instrumental works completed and whole in themselves.

When not performed as part of larger compositions, instrumental music continued to be produced either for court recreation or for practice—as loosening-up exercises designed to improve performance technique. For the former function, the music was variously referred to as *Kapellmeistermusik, Tafelmusik*, or *Hausmusik*. But even here it was not conceived in terms of independent instrumental works. Instead, one finds sets, suites, or *ordres* of instrumental music being published for such purposes. And the same is true of music written for private practice. Many of Bach's collections of instrumental music, notably his 'Goldberg' Variations, were published under the title of 'Keyboard Practice' or 'Exercise' ['*Übung*'], names that did not always even suggest individuation of the pieces constituting the collection. His *Well-Tempered Clavier* was published in 1722 for the 'Use and Profit of the Musical youth Desirous of Learning as well as for the Pastime of those Already Skilled in this Study'. F. W. Marpurg's *Clavierstücke* (1762) was a set of pieces published for pleasurable purposes. In 1756, C. Ph. E. Bach spoke of the publishers who provided collections of 'practical [private] musical art works'.[58]

When Handel published his *Suites de pièces pour le clavecin*, it is said he preferred this French title to the then normal 'Harpsichord Lessons'. Handel's title did suggest individuation between pieces, but he still considered the reasons for publication to be pedagogical. They were also protective. He wrote in a preface to the *Suites*:

I have been obliged to publish Some of the following Lessons, because Surrepticious and incorrect Copies of them had got abroad. I have added several new ones to make the Work more usefull, which if it meets with a favourable Reception; I will still proceed to publish more, reckoning it my duty, with my Small Talent, to serve a Nation, from which I have receiv'd so Generous a Protection.[59]

qui sert de début aux opéra et autres drames lyriques d'une certaine étendue' (*Œuvres complètes de J. J. Rousseau*, 'Beaux-Arts: Dictionnaire de Musique', tr. and ed. V. D. Musset-Pathay (2 vols.; Paris, 1824), ii. 60).

[58] David and Mendel (eds.), *The Bach Reader*, 29 and 269. The term 'Musick' could be used to refer either to one or to many pieces of music.

[59] *The Letters and Writings of Georg Frideric Handel* ed. E. H. Müller (New York, 1935), 11.

Handel did not think, at least on this occasion, that publication or performance of his suites had imposed limits of finality upon them. His lessons were published less because he thought of them as finalized (he speaks of extending them), than for pragmatic reasons having to do with forms of misrepresentation rife in his time. That Handel thought about his music as being misrepresented shows perhaps that expectations were changing, but they had not changed so much that his scores were no longer being adapted and altered by other musicians for their own use, freely and without regard for the original composer's decisions.[60]

Handel's comment is also important because he uses the term 'work' to cover the entire publication, without regard for the number and individual style of each composition or piece contained therein. That was a normal procedure. Before 1800, the terms 'work', '*Werk*', '*Opus*', and '*Partitur*' were often applied to a single publication—to the publication, in other words, of a given amount of music. The terms tended to apply to the music published, rather than to individually formed compositions. Accordingly, the terms also came to refer to a composer's entire output or part thereof. Hence one would speak of a composer's musical opus, work, or *œuvre*.

One hears of a list, for example, being made 'of the Compositions selected from the [published] Works of Handel'. And when 'his very humble and very obedient servant' Bach wrote a letter of dedication to 'his Royal Highness Monseigneur Christian Ludwig', to accompany the manuscript of his *Six concerts avec plusiers instruments* (the Brandenburgs), he arguably indicated something important. Speaking of some 'pieces of my Composition', he perhaps indicated that he thought of them as constituting a single or total compositional work, despite his having constructed and perfected each piece independently of any other. For the phrase 'pieces of my Composition' could have referred to the fact that they were pieces constituting a total composition—a set of pieces—as well as the fact that they were pieces composed *by* Bach.[61] In any event, whether the Brandenburg(s) constitute one or six 'works' is not clear.

[60] In 1724, Couperin added dance suites to his published sonatas of 1692 (Mellors, *François Couperin*, 112). In Forkel's *Johann Sebastian Bach*, the editor C. S. Terry interprets Forkel's general biographical remarks as suggesting that, for Bach, publication postulated excellence, but not necessarily a 'finished work of art' (115).

[61] Cf. David and Mendel (eds.), *The Bach Reader*, 82–3. Apparently, by contrast, Bach considered the B Minor Mass a collection of four compositions, so Friedrich Smend has argued (ibid. 427).

What is clear is that it was not yet normal to use the terms 'work' or 'opus' to denote a single instrumental composition. Instead, any of the following terms seem to have sufficed: '*le morceau detaché*', '*la pièce de musique*', '*die Solo*', '*die Musik*', '*das musikalische Stück*', 'composition', or 'piece'. These terms referred either to a musical or instrumental part or bit of a composition, or to a composition as a whole, depending on the circumstances of performances—how much of the music was played. Other commonly used terms reflected the particular styles or forms within which bits of music were composed. Hence, one finds references to 'aires', 'solos', 'allemandes', 'courantes', 'sarabandes', 'gigues', 'minuets', and 'galanteries', as Bach called the latter. One also finds references to sonatas, symphonies, and concertos. Any published combination of these constituted a work of music.[62]

But no completely *standard* or *generic* way of referring to individual and completed instrumental works yet existed. The absence, of course, merely reflected the lack of a standard and regulative conception of composers producing original, completed, and individually distinct instrumental works. So, even though Rousseau defined 'piece' as an '*ouvrage de musique* of a certain length', which sometimes 'generically' denotes an '*ouverture*' composed in three parts, or 'an *opéra* divisible into acts', he added a proviso. The word '*pièce*', he wrote, is 'very particular' in the domain of instrumental music, when it designates music for 'certain instruments', say, the *viole* or *clavecin*. For one does not speak of a '*pièce de violon*'; one speaks of '*une sonate*'. Yet one speaks of a '*pièce de clavecin*', but not of '*une sonate de clavecin*'.[63]

Only at the end of the eighteenth century did individual instrumental compositions begin to be thought about as self-sufficient works, each publishable in its own right. In 1785, plans were put forward by Samuel Arnold to produce a complete edition of Handel's works. It was completed two years later. Between 1798 and 1806 new editions of

---

[62] Cf. entries relating to the composition and performance of music in S. de Brossard, *Dictionaire de musique* (Paris, 1703) and Walther, *Musicalisches Lexicon*. Also David and Mendel (eds.), *The Bach Reader*, 105, and *Roger North on Music*, pp. xx, 13, and 197.

[63] 'Pièce, s. f. Ouvrage de musique d'une certaine étendue, quelquefois d'un seul morceau, et quelquefois de plusieurs, formant un ensemble et un tout fait pour être exécutée de suite: ainsi une ouverture est une *pièce*, quoique composée de trois morceaux, et une opéra même est une pièce, divisée par actes. Mais outre cette acception générique, le mot *pièce* en a une plus particulière dans la musique instru-mentale, et seulement pour certains instruments, tels que la viole et le clavecin; par exemple, on ne dit point une *pièce de violon*, l'on dit *une sonate*; et l'on dit guère une sonate de clavecin, l'on dit une *pièce*.' (Dictionnaire de Musique', 81–2.)

Mozart's, Haydn's, and Bach's works were produced. In each case, and for the first time, individual compositions, now either vocal or instrumental, were assigned individual opus numbers (when it was possible to do so). With opus numbers individuating works, the need diminished for publishers to refer on the title-pages of their publications to the occasional performance or the purpose for which the music had once been composed. There were other reasons for this change as well, the overall result of which was that if such references continued to be made at all, they were often and increasingly over time relegated to the status of historical footnotes.

It was at this time that the term 'work' ('piece', 'composition', and their foreign equivalents) came generically to signify any original and completed and whole composition of music, whether instrumental or vocal. Thus, in the first, 1880 edition of Grove's *Dictionary of Music and Musicians*, it was correctly observed that the word 'piece' has 'since the end of the last century been applied to instrumental musical compositions as a general and untechnical term'.

XII

Many of the conditions I have described as present in early musical practice were still present after 1800. But something quite significant happened around that date that shifted attention away from a functional view of musical performance, towards a view that focused all its attention on the fact that, amidst all musical activities, works were being produced. This shift in attention changed the way musicians thought about music; it changed their expectations and ideals about the basic conditions of their practice.

Still, it took some time before all activities and procedures were able to fully adapt themselves to the modern way of thinking about music in terms of the production, performance, and preservation of works. That is why one sees considerable continuity across the two centuries. And before any real adaption could take place answers had to be found to those two most basic questions that had emerged out of contemporary musical and aesthetic theory: how could musical works be treated and preserved in a form appropriate to a temporal art of sound, and how could they be preserved to match the romantic, aesthetical beliefs that had come to be associated with them?

The answers musicians gave to these questions in the first half of the nineteenth century will occupy our attention in the next chapter. It will not undermine the argument of that chapter in any way, however, to concede immediately that the theory and practice never succeeded in jelling or matching perfectly—after all, they never had done so before. Nor does it matter that some activities and beliefs continued unaltered from the eighteenth into and through the nineteenth century. And nor does it matter, furthermore, that some practical changes might have occurred for reasons not first articulated in aesthetic and musical theory, or that others did not always occur as quickly or quite in the way musicians wanted them to. What matters, only, is that musical practice did change in its dominant conception around 1800.

How, then, did practice come to reflect a major paradigm shift in the way musicians thought about musical production, and how did the norms of such production each come to be entirely subordinated to the demand that musicians, whatever their activity, be 'true to the work'?

8

After 1800: The Beethoven Paradigm

Franz Liszt thought he had found the perfect way to treat the temporal art of music as a truly fine art. He declared in 1835:

In the name of all musicians, of art, and of social progress, we require: . . . the foundation of an assembly to be held every five years for religious, dramatic, and symphonic music, by which all the works that are considered best in these three categories shall be ceremonially performed every day for a whole month in the Louvre, being afterwards purchased by the government, and published at their expense.

In other words, he continued, 'we require the foundation of a musical Museum'.[1]

Liszt's proclamation had precedents. In 1802, Forkel wrote that 'the most efficacious means of preserving in lasting vigour musical works of art is undoubtedly the public execution of them before a numerous audience.' Public performance of Bach's works (for it was those he was writing about) would 'raise a worthy monument to German art', as well as 'furnish the true Artist with a gallery of the most instructive models'. In 1809 Carl Maria von Weber remarked on the recent foundation of a Museum in Stuttgart. He described it as a meeting-place for professional and amateur artists, which 'seemed to promise well for the development of artistic taste'. Unfortunately, he then added, the Museum had already been reduced in Stuttgart to a 'Reading Society'.[2]

That was in Stuttgart. In other places, works of music had begun to be performed, not every day over a limited period of time, but every so often over a longer period of time. Ideally, it was hoped, these works would be played every so often—forever. In a musical museum replicating the conditions of a museum for the plastic arts, works did not need to be heard every day, as Liszt suggested; it was sufficient that they be on semi-permanent display. They could be heard in

---

[1] 'On the Position of Artists and their Place in Society', in Walker, *Franz Liszt*, 159–60.

[2] David and Mendel (eds.), *The Bach Reader*, 296 and 298; Weber, *Writings on Music*, tr. M. Cooper, ed. J. Warrack (Cambridge, 1981), 34. The term 'museum' was also used to refer to private musical societies.

performances by respectful members of a musical meeting just as often as paintings and sculptures were viewed by gallery visitors. Otherwise they would be stored away, but always under the condition of readiness for exhibition.

Still, Liszt was justified in reiterating a demand for the establishment of a musical museum, for conditions of practice did not change overnight. Weber had already pointed that out. Most of the changes that fostered the emergence of the regulative work-concept spanned many decades. Things had begun to change in significant ways in the 1770s (if not before), numerous changes occurred around 1800, and many if not all the changes stabilized during the course of the nineteenth century. All these changes shared a common aim. They marked a transition in practice, away from seeing music as a means to seeing it as an end. More specifically, they marked a move away from thinking about musical production as comparable to the extra-musical use of a general language that does not presuppose self-sufficiency, uniqueness, or ownership of any given expression. In place of that, musical production was now seen as the use of musical material resulting in complete and discrete, original and fixed, personally owned units. The units were musical works.[3]

I

A crucially important change concerned the social status of composers. As the eighteenth century drew to a close, musicians were no longer thought about predominantly as in service to extra-musical institutions. Like their musical compositions, they were fast being liberated from the traditional power and restraint of ecclesiastical and aristocratic dignitaries. They were also being freed from the demand that they engage, if they could, in theoretical speculation as philosopher-scientists. Instead, musicians—especially the composers amongst them—were sharing in the revolutionary freedom claimed by a rising professional middle class, and gradually, through their liberation, were coming to be seen as independent masters and creators of their art.

---

[3] Charles Rosen employs a similar metaphor when he contrasts artistic and pragmatic uses of language. In art, he notes, we are interested in what is exceptional, not what is normal. Each work sets out on its own to be judged according to its own peculiar merits. 'Individual statement provides the norm and takes precedence over general usage' (*The Classical Style*, 21–2).

In cosmopolitan cities, composers were finding the most fruitful means to escape their former 'social tutelage'. Carl Maria von Weber identified the cosmopolitan composer's hopes, when he noticed, whilst in Prague, that complaints were being heard from resident artists about their circumstances. Such circumstances, he explained,

make it difficult for them to achieve the mentality and the spirit which mark the artist who is a real cosmopolitan and therefore free. Every artist in Prague owes his existence to some noble family and bears the title 'Composer to His Excellency So-and-So.' His opinions are those of his patron, who in his turn champions his own composer against the rest. The result is an absence of the spirit that refuses to be content with merely earning a livelihood and longs to embark on the high seas of art in search of new discoveries.[4]

In these cities, composers believed that they should be able to live and function as free individuals, and that their productive activities should, if they should be subject to anything at all, be subject to the forces of an urban market for music. If they continued to receive patronage from church or court, they did so on the understanding that patrons were not to interfere with or control their creativity. As free persons, they could exercise choice over the form of their patronage, whether it be more or less demanding and restricting—or so they wanted to believe. Whatever social standing accorded most satisfactorily with their aspirations to sail on 'the high seas of art' was the standing they wanted freedom to adopt.

Though it was not until well into the nineteenth century that composers began to be fully accepted as independent persons, that was well after the expectation to act and be treated in this way had become firmly fixed in their minds.[5] Composers of the late eighteenth century, such as Mozart and Haydn, marked a transitionary phase. Weber and Beethoven requested much more explicitly than their immediate predecessors that their social status reflect the romantic descriptions of their new autonomous art.

It was Beethoven, more than any other composer, who set the example for future composers. Throughout his century, his actions had, intentionally and inadvertently, both negative and positive influence. Negatively, he provided others with eyes to see that the

---

[4] Weber, *Writings on Music*, 130.
[5] Cf. Alan Walker's remark that it was not until the 1840s that certain social barriers were broken down so that composers could act, and be fully accepted, as superior beings (*Franz Liszt*, 287).

discontent of the liberated composer could match that of the traditional court musician. Positively, he showed how composers could take artistic advantage of the autonomous art of music. Ultimately, he changed and was believed to have changed so many things having to do with how musicians thought about composition, performance, and reception, that the subsequent Beethoven mania, or the Beethoven Myth as it has come to be called, is justified, if such a thing is ever justified, on much more than aesthetical grounds alone.[6]

For present purposes, Beethoven showed his contemporaries and descendants that modern, liberated composers differed from their predecessors in having a choice as to the source of their livelihood and in being able (in theory at least) to make use of or exploit this choice in whatever ways they saw fit. Monteverdi, Bach, and numerous other composers of their times might have had desires for more independence, desires perhaps recognizable only by hindsight. But it was not until composers were conceptually freed from strict social dependence upon extra-musical bodies with predominantly extra-musical interests that the desire for independence could be both explicitly articulated and realized. What form did this realization take?

II

What seemed to matter most to composers was their freedom from worldly demands. Their romantic role willingly adopted, composers enjoyed describing themselves and each other as divinely inspired creators—even as God-like—whose sole task was to objectify in music something unique and personal and to express something transcendent. Bizet described Beethoven not as a human, but as a God. Samuel Wesley referred to Bach as a 'Saint', a 'Demi-God', and a 'Musical High Priest', and to his masterpieces as the 'Works of our Apollo'. Haydn's spirit was said to penetrate 'the sanctuary of heavenly wisdom'. He 'brought down fire from heaven, to warm and to illuminate . . . earthly hearts . . . to lead them to a sense of the Infinite'. Baini looked back to Palestrina as an early 'amanuensis of God'.[7] Each

---

[6] For discussion of Beethoven mania, see J. Horowitz, *Understanding Toscanini: How he Became an American Culture-God and Helped Create a New Audience for Old Music* (Minneapolis, 1987), 32 ff., Galkin, *History of Orchestral Conducting*, 350, and Dahlhaus, *The Idea of Absolute Music*, tr. R. Lustig (Chicago, 1989), chs. 1 and 8. Horowitz argues that the mania was generated among things partly by nationalist fervour and partly by musicians who owed musical debts to Beethoven.

[7] W. Salmen, 'Social Obligations of the Emancipated Musician in the 19th Century',

of these descriptions acknowledged the composers' authority to express 'higher truths' within their works, an authority now regarded as contingent upon the composers' separation from the ordinary, everyday world.

If composers could have existed without bodily nourishment or support from external sources, they might never again have ventured outside their newly constructed ivory tower. In 1809, the King of Westphalia employed Beethoven with the understanding that it was 'the aspiration and the aim of every true artist to place himself in a position in which he can occupy himself exclusively with the composition of larger works and will not be prevented from doing so by other duties or by economical considerations'.[8] Some years later, Richard Wagner rejected an offer of a 'position' outright—as 'being incompatible with his dignity'.[9]

As with many changes, in their initial stages, attitudes moved from one extreme to another. Having once had negligible freedom, composers under the influence of the romantic aesthetic soon found themselves with too much. Though appreciative of the new, 'charismatic' authority granted to them, many felt profoundly isolated and ignored. 'Great men are notoriously misunderstood in their lifetime,' Weber lamented:

All too often they themselves die of hunger, only to be applauded after their death by hungry publishers. In fact human beings have a way of neglecting the genius on their doorsteps and appreciating only those who are dead.[10]

Composers thus began to seek a middle position between that of isolated genius and that of Kapellmeister. By compromising some of their freedom, they hoped to live within rather than above society (as if the latter were ever a real possibility). They did not find their search an easy one.

One strategy was to play a double musical role. Some composers wrote two kinds of music, one for a small, 'educated' public capable of appreciating the highest expressions of musical art, the other for a

in Salmen (ed.), *The Social Status of the Professional Musician from the Middle Ages to the Nineteenth Century*, tr. H. Kaufman and B. Reisner (New York, 1983), 269; *The Wesley Bach Letters*, 1, 29, 6; Haydn, *The Collected Correspondence*, 226; W. D. Allen, *Philosophies of Music History: A Study of General Histories of Music, 1600–1960* (New York, 1962), 87; Salmen, 'Social Obligations', 287. I have benefited from Salmen's text in formulating this argument.

[8] *Beethoven: Letters, Journals and Conversations*, ed. M. Hamburger (New York, 1951), 72. Recall, from my Introduction, Hoffmann's plea for poor oppressed composers to be given the means to leave the neighbourhood of windbags and bores.

[9] Salmen, 'Social Obligations', 268.   [10] Weber, *Writings on Music*, 43.

large, unrefined, popular audience. This strategy generated a tension for those caught between composing 'civilized' versus 'popular' forms of 'classical' music. Forkel noticed the tension straightaway.

[A] rock upon which genius often comes to grief, is the public's undiscriminating applause. To be sure many artists have been thrown off their balance by exaggerated and often unmerited plaudits. . . . The public merely asks for what it can understand, whereas the true artist ought to aim at an achievement which cannot be measured by popular standards. How, then, can popular applause be reconciled with the true artist's aspirations towards the ideal?[11]

Some composers, Haydn being the best-known example, though he was also an early case, chose after a brief spell in the cosmopolitan world to return to service at court. 'How sweet this bit of freedom really is,' Haydn had written whilst away from court; 'I had a kind Prince, but sometimes I was forced to be dependent upon base souls. I often sighed for release, and now I have it in some measure. . . . The realization that I am no bond-servant makes ample amends for all my toils'.[12] Apparently his optimism and relief did not last very long.

Some other composers were too well aware of the plight or limitations of servitude ever to return to it. So they tried instead to survive on the basis of public engagements and commissions issued by professional musical bodies, such as the recently formed orchestral societies. Beethoven was one of the first if not the very first composer to receive a commission from a musical organization. It was from the Philharmonic Society of London, which, in 1822, requested from him a symphony—his ninth.

Trying to live on musical commissions and public engagements was extremely difficult. As part of an infant class, composers were subject to new marketing constraints and unstable social forces of which they had no previous experience. However musical works and their composers were being conceived 'in the books', they required publishing houses, performing bodies, and a paying public to sustain their worldly existence. But most of these institutions were newly born themselves, and were therefore unable or unwilling as yet to treat composers in a consistent and fair manner. Beethoven complained:

The manner in which Art is handled nowadays in inexcusable. I have to pay one third of my receipts to the management and one fifth to the prison budget.

---

[11] *Johann Sebastian Bach*, 149–50.
[12] Haydn, *The Collected Correspondence*, 118.

The devil take it! Before it is all over, I shall be asking whether music is a free Art or not.[13]

Beethoven pleaded for the opening of a 'single Art Exchange . . . to which the artist would simply send his works, and be given in return as much as he needs'.[14]

III

Composers never wished to detach themselves from the world completely. If theories of transcendent and formalist emancipation of the early nineteenth century gave composers the theoretical terms to disassociate themselves from the world altogether, social emancipation determined that they position themselves in the world according to their new, independent, and individual consciences.

Some took the purist's route, luxuriating as far as they could in their liberation from 'worldly' constraints, devoting themselves thereby to the composing and performing of a pure sort of music. Beethoven felt and wrote of such desires, though not without conflict. He, like many others, found that he was unable to cut himself off completely from the ordinary world of the extra-musical. While 'purist' ideals prompted him to seek in his music detachment from the ordinary world, democratic ideals prompted him also to seek in his music political freedom within this world for himself and others. He had, he recognized, a concern for 'needy humanity'.

Beethoven was not alone in this conflict. In the early nineteenth century, artists generally felt allegiance to two ideals: 'art for art's sake' and 'art for the people'. Surely conflict would arise in anyone who tried to meet both at the same time. Apparently not. When Goethe wrote that one 'cannot escape the world more certainly than through art and [one] cannot bind [oneself] to it more certainly than through art', he pointed to the fact that in art one meets both ideals, for art has double-sided autonomy.[15]

Any comparable conflict that emerged in the musical world was

[13] From a conversation between J. W. Tomaschek and Beethoven (1814), from O. G. Sonneck (ed.), *Beethoven: Impressions by his Contemporaries* (New York, 1954), 104.

[14] Letter of 1801, *Beethoven*, ed. Hamburger, 35.

[15] J. W. von Goethe, *Elective Affinities*, tr. R. J. Hollingdale (Harmondsworth, Middx., 1971), 196. On one interpretation, the ideals respectively indicated an art produced for the élite and one produced democratically for the masses. But this is not their only interpretation.

resolved as well. Writing on Gottfried Weber's *Te Deum*, Carl Maria von Weber pointed out that no critic could ignore the dedication of the piece to 'Germany's victorious army'. But then he worried: had the composer produced a work 'remarkable more for its suitability to the occasion than for absolute artistic value'? Weber simply asserted that it was a great achievement for a composer to succeed in producing either sort of work. If Weber did not meet the conflict head on, Liszt did. Asking whether a composer should 'be reproached with having given way to his genius rather than to the requirements of the worship', he answered that, in his music, the religious and the revolutionary had just become the purely musical.[16]

Composers found another resolution when they began to describe their music either as absolute or as programmatic. Their descriptions rested heavily on the romantic terms that had given instrumental music its autonomy. In neither sort of music did meaning derive from service to an extra-musical occasion. In both, meaning was achieved through the unique qualities of instrumental sound, though whether the music embodied 'transcendent' content of a spiritual, religious, or even a political sort was believed to make all the difference.

Complicating the story of these types of music is the fact that they each received numerous different descriptions. The most common is given in terms of the distinction between the musical and extra-musical. Thus, absolute music has often been described as purely instrumental and expressive of nothing but a musical idea; programme music, by contrast, as instrumental music whose formal development is guided by, or implicitly refers to, an extra-musical idea which might be political, religious, or poetic.[17] But, articulated in terms of the musical/extra-musical distinction, the concepts of absolute and programme music have always been regarded as profoundly problematic when it came to describing the extent of the difference between them.

How, for instance, can one tell to what belongs the musical and to what the extra-musical? Is Beethoven's Pastoral Symphony a work of absolute or programme music? Were his sentiments about the countryside embodied in the Symphony at some point in the creative process transformed from extra-musical sentiments into musical ones, or did

---

[16] Weber, *Writings on Music*, 127; P. Merrick, *Revolution and Religion in the Music of Liszt* (Cambridge, 1987), 100 and 156.

[17] Wagner coined the term 'absolute music' and Liszt the term 'programme music'. The concepts were present, however, before they were named in this way. See Scruton, *The Aesthetic Understanding*, 37–48, and Dahlhaus, *The Idea of Absolute Music*, ch. 10.

they remain extra-musical? Beethoven provided notes at the Symphony's first performance in which he contrasted modern musical expression with traditional notions of imitation. The Symphony, he wrote, is 'not a painting in sounds', but 'a record of sentiments' evoked by the composer's 'reminiscence of country life'.[18] Unfortunately, he left it ambiguous, supposing that he was even thinking in these terms, whether descriptions of pastoral sentiments are constitutive of literal descriptions of the music, whether they are metaphorical descriptions, or whether, finally, they are just to be treated as aids provided by composers to help listeners comprehend the music's expressive, musical form. Each alternative has been considered in the literature.[19]

The fact that composers did not conceive of the difference between absolute and programme music only in terms of the musical/extra-musical distinction is crucial. For it means that the problems of the latter distinction could not automatically or simply be transferred to the former. In fact, the concepts of absolute and programme music did not emerge so much because composers wanted to classify their work either as one or the other, but more because they wanted their music to be purely musical and religiously or spiritually meaningful at the same time. Some believed this could be achieved through absolute music, others through programme music. Composers were not working so much with a distinction as with competing concepts.

Whether one composed music under the description of absolute or programme music, composers still believed that being faithful to the musical medium was crucial. When, therefore, it was suggested that music had a programme, one did not discover this fact through the music alone. Despite the presence of a programme, the music still consisted only of sounds and thus retained a purely musical meaning. That it also had a programmatic meaning was usually indicated through a composer's use of titles, prefaces, programme notes, or public proclamations. Announcing a work as programmatic did not, however, render its musical meaning impure. That a work could have purely musical and programmatic meaning simultaneously was explained by appeal to the 'transcendent' move. If composers described their music

[18] 'The Sketch-books: Notes on the Pastoral Symphony', rpt. in *Beethoven*, ed. Hamburger, 68.
[19] Recall Liszt's suggestion, for example, that programme music is just that kind of instrumental music where a preface is added 'by means of which the composer intends to guard the listener against a wrong poetical idea of the whole or of a part of it'. Cf. R. Scruton's 'Programme Music', *The New Grove Dictionary*, xv. 283–7.

as programmatic, they could still claim that its meaning was transcendent and, on this level, reconcilable with the demands of formalism. Liszt once claimed that his 'Revolutionary' Symphony represented not just 'a particular battle', but the 'glorious feelings aroused by revolution in general'.[20] Other composers found they could achieve the same reconciliation in absolute music. Thus, Beethoven's Ninth Symphony has become the paradigmatic example of a work of music whose absolute status is not affected by the 'programmatic' content of its 'Ode to Joy'.[21]

Music could thus serve its purely musical and political roles simultaneously under its description either as absolute or programmatic. Where this fact might have motivated composers to give up the distinction, it did not. Composers found it necessary to take sides. The fact that both concepts captured the same function complicated their decision—though not in all cases. Schumann was quite sure of his decision. 'If a composer', he wrote, 'shows me his composition along with . . . a programme, I say: "First show me that you can make beautiful music; after that I may well like your programme" '.[22]

---

[20] Walker, *Franz Liszt*, 114. Cf. Wagner's comment that 'Beethoven followed Haydn's example, but instead of using these popular melodies to provide entertainment to a nobleman's dinner-table, he transmuted them into an ideal sense for the common man,' quoted in B. Bujić (ed.), *Music in European Thought, 1851–1912*, 66. For another comment on this, see J. Barzun, *Berlioz and his Century: An Introduction to the Age of Romanticism* (Chicago, 1982), 190–1.

[21] It is said that Beethoven's Ninth Symphony was composed as 'instrumental music' even in the Finale. Cf. Rosen, *The Classical Style*, 440, L. Treitler, *Music and the Historical Imagination* (Cambridge, Mass., 1989), 25; also Debussy's comment that '[i]t has been stated elsewhere that Beethoven made the spoken word the "apotheosis and crowning glory" of this edifice in sound, and that the finale of the Ninth thus prepared the way for music drama. Isn't that just a convenient theory for the Wagnerians to hold? The intervention of the thousand voices is really to salute the art of music above all else' (*Debussy on Music*, ed. R. Langham Smith (Ithaca, 1977), 15). Finally, in a recent article 'Putting the Program back in Program Music' (*The New York Times*, 3 Sept. 1989, 'Arts Section', 17), Owen Jander defines 'program music' 'in the broadest sense of the term, [as] instrumental music that despite the lack of text, is meant to convey some extramusical idea. In the narrowest sense, it is instrumental music that, like Berlioz's "Symphonie Fantastique," reflects an elaborate descriptive text actually provided by the composer. The historical spectrum, however, is vast: from Bach . . . [to] Tchaikowsky . . . Beethoven came at midroute in this historical journey. Did he compose program music? According to the nearsighted notions of 1905, no, not yet. From the viewpoint of 1805, however, this composer was on the front edge.'

[22] 'On Louis Spohr's Sixth and Seventh Symphonies', *Neue Zeitschrift* (1843), in *Schumann on Music: A Selection from the Writings*, tr. and ed. H. Pleasants (New York, 1965), 184. The same sentiment was later expressed by conductor Wilhelm Furtwängler when he remarked that, where others hear the Napoleonic messages in Beethoven, he hears Allegro con Brio (Horowitz, *Understanding Toscanini*, 102).

Mahler was less sure and wavered constantly between the two concepts. Having admitted being inspired by Nietzsche's *Die fröhliche Wissenschaft* when composing his Third Symphony, he could have claimed the result to be a work either of absolute or of programme music. He chose the former, suggesting on this occasion that, once created, the work can break free from its inspirational source or origin, as a house loses its scaffolding once it has been built. On other occasions, he saw the advantages of his music's being called programmatic. He wrote:

Just as I find it trite to invent music to fit a programme, in the same way I find it unsatisfactory and fruitless to try to invent a programme for a musical work. The fact that the first impulse toward a musical creation has been a concrete, definable experience lived by its composer does not alter this at all. Here we have reached, and of this I am certain, the great form where the two paths of symphonic and dramatic music diverge.[23]

Mahler's last comment complicates the entire story. By distinguishing between symphonic and dramatic music he pointed to the fact that the conceptual development of programme music had depended not only on developments in absolute and instrumental music, as I have suggested so far, but also on those in 'dramatic' music—notably in opera.

Before 1800, opera had mostly been conceived as a secular art of popular, dramatic entertainment, or as a religious form of composition designed for serious audiences. Early thinkers placed emphasis and value on the dramatic text over and above the sonic structure. In this respect they saw opera as they did all other productions involving music. Ridiculing operas performed in London in a foreign language, Joseph Addison pointed not only to growing nationalistic sentiments in the Europe of the early eighteenth century, but also to the importance of an audience's comprehending the dramatic text. He surmised:

Our grandchildren will be very curious to know the reason why their forefathers used to sit together like an audience of foreigners in their own

---

[23] Quoted in H.-L. de la Grange, *Mahler*, i (New York, 1973), 357. Cf. Bruno Walter's thoughts on this matter. Nature itself seems to be *transformed* into sound, he begins. The movements follow a predetermined sequence of ideas. Having given titles to each of the movements we see the basic structural unity of the Symphony. Given this we can do without the titles. Titles can be dropped like a scaffolding when the house is ready. The Symphony becomes pure music (*Gustav Mahler*, tr. J. Galton (New York, 1970), 112). In *Der Anzeiger*, a critic wrote of the same symphony: 'With or without a program, the composer wants us to listen to his tone-poem as pure music, and this is possible thanks to the expressive quality of his themes and the clarity of his construction' (de le Grange, *Mahler*, i. 392).

country, and to hear whole plays [and operas] acted before them in a tongue which they did not understand.[24]

Modern opera has not been plagued by this demand for immediate intelligibility either to the same extent or for the same reasons. And this is not just because translation has become easily available in accompanying programme notes or in electronic subtitles. In 1955, the London *Observer* reported that Sir Edward Appleton did not mind what language an opera was sung in, so long as it was sung in a language he did not understand.[25] This view, however odd, is not uncommon. For, since the end of the eighteenth century, opera has come to be conceived in either of two ways: first, as an art-form that synthesizes in Wagnerian style all the arts to form an undifferentiated unity of the sort that existed in Antiquity; second, as a derivative or hybrid form of a specifically musical art.[26]

Seen as the latter, opera does not stand opposed to purely instrumental music, but as a hybrid form of it. Attention is focused less on the drama and text, as it was in early operas, than on the transcendent language of operatic music. No one was more influential in bringing about this shift in attention than Mozart, of whom it was said that

the operose character of [his] accompaniments was long made an objection to his dramatic music. . . . The Emperor Napoleon once inquired of the celebrated Grétry, what . . . the difference was between Mozart and Cimarosa. 'Sire', replied Grétry, 'Cimarosa places the statue on the stage, and the

[24] *The Spectator*, 21 Mar. 1711. Addison was discussing the progress of Italian opera on the English stage. He remarked particularly on the very poor translations provided for the Italian texts.

[25] From I. Crofton and D. Fraser (eds.), *A Dictionary of Musical Quotations* (New York, 1985), 105. Cf. Stendhal's remark: 'As you may have guessed, nobody gives a damn for the librettist; for who indeed . . . would dream of judging an opera by the *words*?' (*The Life of Rossini*, 147). Recall, as well, Schoenberg's discovery that reading the poems for Schubert's Lieder made no difference to his comprehension of the latter ('The Relationship to the Text', *Style and Idea: Selected Writings of Arnold Schoenberg*, ed. L. Stein, tr. L. Black (Berkeley/Los Angeles, 1975), 144).

[26] In its time, Wagner's theory of the total work of art (*Gesamtkunstwerk*) threatened neither the regulative work-concept nor the belief in aesthetic autonomy. His theory only threatened the recently evolved classification and differentiation of the fine arts. In fact, some have viewed Wagnerian opera as the culminating point in the development of the concept of the work of art, in which all emancipated arts find their proper place and function in the aesthetic and artistic domain. For discussion of this view, see M. Tanner, 'The Total Work of Art', in P. Burbidge and R. Sutton (eds.), *The Wagner Companion* (London, 1979), 140–224, and for historical background on conceptions of opera, see P. Kivy, *Osmin's Rage: Philosophical Reflections on Opera, Drama, and Text* (Princeton, 1988). Kivy favours viewing 18th-cent. operas also as constitutive of an essentially musical genre.

pedestal in the orchestra; while Mozart puts the statue in the orchestra and the pedestal on the stage'.[27]

Just as opera came to be seen itself as a hybrid from of absolute music, so programme music did also. In both cases, extra-musical components were not believed to undermine the essentially musical nature of the compositions. In the contrasting view, however, inasmuch as opera was still regarded as much a dramatic as a musical art, programme music could be viewed either as before, as a musical art, and thus different from opera. Or, now, it could be viewed as an art-form that, like opera, had certain extra-musical elements that could not be ignored or completely subordinated to the musical content. Many felt threatened by this second understanding of programme music. Hanslick, for example, was quite suspicious of any music that had extra-musical components, on the grounds that the latter were necessarily 'extraneous' to the beauty of music which, in his view, was found solely in the tonally moving forms. He therefore concluded that Liszt's programmatic symphonies 'succeeded more completely than anything heretofore in getting rid of the autonomous significance of music'.[28] Hanslick had misunderstood Liszt's programme music, for it does not ever seem to have been the latter's aim to bring about such a consequence.

The debate over absolute and programme music was clearly complicated by its association with the distinction between the musical and extra-musical. For, with this connection, musicians feared that the autonomy so recently granted to music would be undermined. In this respect, the debate between supporters of absolute music and those of programme music—more than a distinction as such between two different types of music—reveals just how important it had become to composers that they be able to produce music that could say something about the world without that admission compromising

---

[27] Hogarth, *Music History*, 78. Berlioz was less than happy with this view of things. In an essay on Beethoven's *Fidelio* (*A travers chant*, 73) he writes: 'Ce qui ne veut point dire que cette partie [la partie vocale] ne soit pas restée prédominante, ainsie que le prétendent les éternals rabâcheurs du reproche addressé par Grétry à Mozart: "Il a mis le piédestal sur la scène et la statue dans l'orchestre," reproche fait auparavant à Gluck, et plus tard à Weber, à Spontini, à Beethoven, et qui sera toujours fait à quiconque s'abstiendra d'écrire des platitudes pour la voix et donnera à l'orchestre un rôle intéressant, quelle que soit sa savante réserve.'

[28] Reported by Payzant in his Introduction to Hanslick's *On the Musically Beautiful*, p. xxiii.

the independence of their musical medium. As I said, some composers found a resolution in absolute music, others in programme music.[29]

IV

Looking back to a time when composers produced occasional music, the conclusion drawn in hindsight was that works of music 'can rarely be composed under a contract of service, since such a contract involves such immediate control over the labours of the servant by the employer', that it is at odds with 'the preparation of a work of art'.[30] Precisely so. When composers began to view their compositions as ends in themselves, they began to individuate them accordingly. When composers began to individuate works as embodied expressions and products of their activities, they were quickly persuaded that that fact generated a right of ownership of those works to themselves. Thus, as music came to be seen as the product of a free person's labour, a change was deemed necessary in ownership rights.[31]

In the early eighteenth century, publishing houses acquired copyright over music, at least insofar as sheets of music were produced. For most of the eighteenth century copyright remained so defined. In 1793, however, copyright laws were passed in France to transfer ownership away from publishers to composers. Germany and England followed suit some years later. Though the new laws were not universally accepted, the rationale behind them was clear. The new laws reflected the basic idea that composers are the first owners of their works, for it is they who put the works in permanent form.[32]

A report by Jacques Attali provides us with a further sense of what changes in copyright involved. His report concerns a regulation passed in France in 1786 by the *Conseil du Roi*. The *Conseil* noted that ' "the piracy of which the composers and merchants of music were complaining was injurious to the rights of artists . . . , and ownership rights were daily becoming less respected, and the talented were

---

[29] A significant part of the threat to music's new-found autonomy came from the rife censorship policies of 19th-cent. Europe. I have discussed this threat in detail in my 'Music Has no Meaning to Speak of: On the Politics of Musical Interpretation', forthcoming in M. Krausz (ed.), *The Interpretation of Music* (Oxford). See also H. Raynor, *Music and Society since 1815* (London, 1976), ch. 1.

[30] 'Copyright', Grove, *Dictionary*, ii. 431.

[31] Cf. Hegel's discussion of abstract right, copyright, and plagiarism, in his *Philosophy of Right*, tr. T. M. Knox (Oxford, 1967), 54–7.

[32] See 'Copyright', Grove *Dictionary*, ii. 430; also 'Copyright', *Encyclopedia Britannica*, 11th edn. (Cambridge, 1910), vii. 124.

deprived of their productions" '. The *Conseil* resolved, therefore, that the privilege of the seal should henceforth be granted to commercial publishers, after they had 'justified the transfer of rights . . . made to them by the authors and owners'. 'The regulation', says Attali, 'also specified the form and terms of the declarations and registration necessary to assure ownership rights.' Lastly, it prohibited ' "under penalty of a fine of 3,000 livres," the unauthorized use of any piece of music as well as engraver's stamps and trademarks.' In this regulation, Attali concludes, 'the ownership right of authors over their "works" was finally recognized'.[33]

In 1842, British copyright laws were revised to grant rights to composers over the performances of their works as well as their scores. It is noteworthy that the new laws still extended copyright to scores and performances but not directly to the works themselves, even though granting copyright in this way presupposed the existence of musical works conceived as products that were owned and marketed. The intimate relation between the elusive, abstract presence of works and their concrete scores and performances was, however, recognized in the British Copyright Act of 1911, where a distinction is drawn between the abstract property of the copyright and the material (concrete) objects representing it.

The kind of protection afforded by copyright laws in the nineteenth century did not extend to all forms of music but only to original musical works. Popular and folk musics were initially excluded, and that, Attali notes, caused quite an uproar in France. Though the word 'original' was defined not in terms of artistic originality, but, rather, in terms of independent labour and skill, the initial intention still seems to have been to define copyright according to the conception of a 'civilized' work of art. That 'civilized' music in this period was the only form of music adequately or fully notated might explain the exclusion. Maybe it does not. When it was suggested mid-century that copyright laws, royalties, taxes, and the like, be instituted in the production of popular music, it was objected that this sort of music was artistically and economically worthless. In 1850, *La France musicale* announced with typical prejudice that 'if you create operas, symphonies, in a word, works that make a mark, then royalties shall be yours; but taxing light songs and ballads, that is the height of absurdity'.[34]

---

[33] *Noise*, 52.   [34] Ibid. 78.

That originality was understood for copyright purposes in terms of independent labour and skill had an interesting consequence. Under the law, what today we would consider a single work scored for two different sets of instruments, say, either for piano or for full orchestra, constituted two separate works of music, each of which was subject to its own copyright protection. This followed from the fact that each scoring entailed independent labour and skill. In theory, copyright laws also allowed two separate composers to produce near identical scores.[35] But I say 'in theory' because, just when copyright laws were passed to protect composers, a new understanding of plagiarism acquired force enough to prevent the public acceptance of such eventualities.

V

According to the *Oxford English Dictionary*, the English term 'plagiarism' was first applied to music in the *Monthly Magazine* of 1797. There a certain song was said to be 'a most flagrant plagiarism from Handel'. Though the concept of plagiarism was already in use before this time, its early use had been complicated by the fact that there had been a rather different conception of what it meant to compose music, a conception that had allowed for substantial borrowing among composers of their musical materials. When, however, the ideal of originality began to regulate the activities of composers, a corresponding change was required in the concept of plagiarism.

The demand for originality translated into a demand that each composer should create his works from scratch. 'The singular and original seemed to be his chief aim in composition,' Tomaschek reported of his contemporary, Beethoven, citing for evidence an answer Beethoven once gave 'to a lady who asked him if he often attended Mozart's operas. "I do not know them," he replied, "and do not care to hear the music of others lest I forfeit some of my originality." '[36] Beethoven's comment is largely polemical, since he, like every composer who followed him, studied carefully and allowed himself to be influenced by the music of the past. Notwithstanding, the ideal of originality Beethoven endorsed had new and central significance, as Claude Debussy later explained.

---

[35] 'Copyright', Grove, *Dictionary*, ii. 430.
[36] 'Johann Wenzel Tomaschek' (1798), in Sonneck (ed.), *Beethoven*, 22.

Beethoven's real lesson to us was not that we should preserve age-old forms, nor even that we should plant our footsteps where he first trod. We should look through open windows into clear skies. Many people appear to have closed them, seemingly for good; those successful so-called geniuses should have no excuse for their academic contrapuntal exercises, which are called (out of habit) 'symphonies'.[37]

When Mozart recomposed his C major Oboe Concerto for another instrument, he was gently criticized for not having produced an original work. Rossini was later harshly criticized (mostly by his enemies) for writing the same music too many times. The repeated use of themes and passages in different works of music was increasingly regarded as lazy and unworthy. The familiar criticism that began to develop in the early nineteenth century, that Vivaldi had composed the same concerto 300 times, carried weight only from the modern but not from the early perspective. It was, of course, a blatantly anachronistic accusation.[38]

The ideal of originality had an immediate impact on how composers viewed their compositional procedure. It helped them follow other artists, for example, in dispensing with rule-following models of composition. Transcending rules was just the sort of thing composers found necessary to help guarantee originality in their compositions. In a telling encounter with Beethoven, Ferdinand Ries recalled that he had once

mentioned two perfect fifths, which stand out by their beauty of sound in one of his earlier violin quartets, in C minor. Beethoven did not know of them and insisted it was wrong to call them fifths. . . . When [however] he saw I was right he said: 'Well, and who has forbidden them?' . . . I replied: 'It is one of the fundamental rules.' Again he repeated his question, whereupon I said: 'Marpurg, Kirnberger, Fuchs, etc., etc., all the theoreticians!' 'And *so I allow them!*' was his answer.[39]

---

[37] *Debussy on Music*, 15–16.

[38] Recall Stravinsky's remark that Vivaldi is 'much overrated—a dull fellow who could compose the same form over and so many times over', in I. Stravinsky and R. Craft, *Conversations with Igor Stravinsky* (Garden City, New York, 1959), 84. For discussion of Rossini's reuse of materials, see Stendhal, *The Life of Rossini*, 161 and 400–1. As Dahlhaus also points out in *Nineteenth-Century Music*, Rossini might have defended himself on the grounds of his not being involved in the production of high art, and therefore not subject to the evaluative criteria of such (p. 8).

[39] 'Ferdinand Ries' (1801–5), in Sonneck (ed.), *Beethoven*, 49. Apparently some messages need constant reiteration. Erik Satie expressed the same attitude when he wrote that 'In art there must be no slavery. With each new composition, I have tried to baffle any followers I may have had, as regards both the form and content of my works.

Allying themselves again with all creators of fine art, composers began to conceive of their works as discrete, perfectly formed, and completed products. Music soon acquired a kind of *untouchability* which, translated into concrete terms, meant that persons could no longer tamper with composer's works. The demand that one's works be left alone was rationalized according to the romantic belief that the internal form and content of each such work was inextricably unified, or by the belief that works were specified *in toto* according to an underlying or transcendent truth. That a work's determining idea was an expession of an individually inspired genius effectively meant that its content was necessarily elusive and not subject, therefore, to mundane description or change. That being so, the practical outcome was to instil fear in those who dared to touch a work, on the grounds that they would probably damage it irreparably and forever.[40]

Demands for originality and untouchability put a stop not only to tampering with works, but also to the open borrowing of music. For how could one be original if one used another's music? To accommodate, however, a continuing interest in using pre-composed music, a distinction emerged to differentiate two sorts of composition, 'original' and 'derivative' composition.[41] In addition to the composition of original works, new activities emerged conceived under the banner of composing versions of pre-existing works. Transcription, orchestration, and arrangement were the names given to such activities. Each was described as being bound and limited by the presence of an already

Only thus can an artist avoid becoming the leader of a "school"—that is to say a pundit.' P. D. Templier, *Erik Satie* tr. E. L. and D. S. French (Cambridge, Mass., 1969), 44.) And just a few years ago, Philip Glass recalled a time when he noticed that he had 'been operating under a lot of rules that had become automatic, and that there were things that weren't possible to do in [his] music because [he] had made them forbidden'. He said to himself ' "Why can't I do it?" "Well, there's the rule." "Rule!!?" "Who's making the rules?" "I'm making the rules." "And that was the end of the rule." ' (Quoted in 'Philip Glass', in J. Rockwell, *All American Music: Composition in the Late Twentieth Century* (New York, 1983), 120.) Note, also, that in his most recent book Leonard Meyer usefully distinguishes two notions of originality, one whereby a composer formulates new rules; the other whereby a composer devises a new strategy for using already existing rules (*Style and Music*, 34).

[40] Sometimes even the original composer instilled himself with such fear. Beethoven once remarked that he was not accustomed to revising (*retoucher*) his compositions once they were finished. 'I have never done this,' he continued, 'because even the slightest change alters the character of the composition' (Letter (in French) of 1813, in *Letters of Beethoven*, ed. E. Anderson (3 vols.; New York, 1961), i. 405).

[41] Cf. Warren Dwight Allen's comment that 'there is no sonata or symphony . . . which has been passed on for completion or modification to successive generations; each composer writes his own' (*Philosophies of Music History*, 301).

existing work and, therefore, as not being strictly creative. Thus, when Johann Adam Hiller orchestrated some of Handel's compositions in 1798, it was expressly stated that he would try to retain the essence of the original.[42] When Liszt began to transcribe music, he sought, as we already know, to unfold the original works as accurately as possible.

Thus, though it remained admissible after 1800 to 'recompose' music, that activity was now understood in terms of producing versions or variants of works. It was no longer allowable for someone to complete or recompose an original work of a given composer. Better to have an unfinished symphony regarded as authentic and even finished (à la Schubert's B Minor Symphony of 1822), than to have a symphony inauthentically finished or completed through compositional activity distinct from the original one. But the fact that new compositional procedures did not dispense altogether with the use of pre-composed materials meant that new policies had to be decided upon to determine the limits of such use.

When one transcribed or arranged works, even when one quoted a piece of music composed by someone else or by one's earlier self, or, finally, when one chose to compose 'in the style of', it was expected that one publicly acknowledge that fact. If one tried to deceive the public, by claiming to have authored any or all parts of the composition, one would be accused accordingly. Nothing had changed regarding attitudes towards deception as such. What changed was the degree of freedom given to musicians to use previously composed music. And new policies—plagiarism policies—were introduced to accommodate that change.

Plagiarism policies were introduced, therefore, to determine what counted as a purloined idea and what counted as theft, be it of an idea, a design, or a large passage of a work. More than that, they were designed to grant composers more 'artistic' protection of their works than copyright laws did at first by themselves. Gradually they were incorporated into the copyright laws. Now, if one composer wrote a work for piano and another wrote one for orchestra, and it was found that the compositions were wholly or partially identical apart from differing instrumental specifications, procedures existed by which to determine whether one work involved theft of the other.

---

[42] Hogwood, *Handel*, 246.

VI

When composers began to request that *their* notational instructions for the performance of *their* music be followed to the mark, they were asserting their new authority in the strongest way they knew how. What they were demanding in fact was the translation of the ideal of untouchability into concrete terms. If a work was untouchable, then barring obvious extenuating circumstances so was its representation by the composer in notational form. The notational form in which composers gave their works to their publishers, or anyone else, was the form in which their works were to remain.

Haydn thus requested of Artaria and Company that they should not

let anyone copy, sing, or in any way alter these *Lieder* before publication, because when they are ready I will sing them myself in the critical houses. By his presence, and through the proper execution, the master must maintain his rights: these are only songs, but they are not the street songs of [Leopold] Hofmann, wherein neither ideas, expression nor, much less, melody appear.[43]

Beethoven also wrote to his publishers, to request even more stringently that they should not tamper with his works. 'Why did you not print the date, year, month and day, as I wrote them on my score?', he asked: 'You will give me a written assurance that in future you will preserve headings just as I wrote them.'[44] In 1805 a reviewer for the *Sun* felt obliged to record his protest 'against any such alterations in works [of Handel and Mozart] that have obtained the sanction of time and of the best musical judges'.[45]

Notation had to undergo significant changes in function and form before composers could demand that their scores should not be tampered with. In the last half of the eighteenth century, composers were increasingly allowed to produce completed scores to reflect the very best version of their music, conceived independently from any particular performance. The need for a fully specifying notation really became urgent, however, when it became the norm for music to travel

[43] 20 July 1781, *The Collected Correspondence*, 31.

[44] *Beethoven*, ed. Hamburger, 101.

[45] Hogwood, *Handel*, 247. Wesley complained bitterly of the 'diabolical' and 'miserably mangled and mutilated' scores of Bach's works that were being passed around, and of his consequent intention to secure the publication of an authentic edition of them all (*The Wesley Bach Letters*, 22, 4, and 7).

independently of the composer, when one and the same composition began to be repeated in numerous performances, when compositional styles became more personal, and when, finally, musicians who had no personal contact with composers fully realized their need for some intelligible and accurate means of access to their music. Consequent developments in notation included increased specificity of structural elements, standardized symbolism, and improved copying. Such developments were designed to demonstrate the ways in which music could now be preserved in a manner suitable for a fine art.[46]

In 1799, in an announcement soliciting subscriptions to Haydn's *Creation*, the composer expressed pride over his new work that

> is to appear in three or four months, neatly and correctly engraved and printed on good paper, with German and English texts; and in full score, so that, on the one hand, the public may have the work in its entirety, and so that the connoisseur may see it *in toto* and thus better judge it; while, on the other, it will be easier to prepare the parts, should one wish to perform the work anywhere.[47]

Beethoven also demanded diligent readings of his scores. Influential in replacing vague tempi markings such as 'andante' and 'adagio' with precise metronomic markings, he was eager to make the replacement well known. He wrote to Schott's and Sons:

> The metronome marks will follow soon, do not fail to wait for them. In our century things of this kind are certainly needed. Also I learn from letters . . . that the first performance of the symphony received enthusiastic applause, which I ascribe mainly to the use of the metronome. It is almost impossible now to preserve the *tempi ordinari*; instead, the performers must now obey the ideas of the unfettered genius.[48]

Beethoven wrote this letter in 1826. Long before that, however, metronomes had been used in performances in France and England by conductors who found that their time-beating did not always keep an ensemble together. But these early uses of the metronome reflected an interest in a musical ensemble's keeping time, more than an

---

[46] According to Adam Carse, the first full scores of symphonies to be printed were Haydn's symphonies published by Leduc in 1801 (*The Orchestra in the Eighteenth Century* (Cambridge, 1940), 117).

[47] Haydn, *The Collected Correspondence*, 155.

[48] *Beethoven*, ed. Hamburger, 254.

interest in the specificity of scores. Thus, for example, Roger North wrote in the early eighteenth century:

In all time keeping . . . an artist will, with the agency of his hand, not onely shew the gross beats, but also distinguish to the eye, crochetts, and even quavers . . . which is a vast help to keep the performers together.[49]

The idea of keeping time was a feature both of early and modern performance practice. No significant change occurred with regard to this, except that orchestras, as they were 'professionalized' at the end of the eighteenth century, were required to perform with greater skill, care, and precision than ever before.

The significant change around 1800 was the extent to which metronomic markings expressed a new interest in producing scores containing as few ambiguous marks as possible. Ambiguous markings had traditionally misled performers or encouraged them to play extempore and not yet in a manner devotedly compliant with a composer's instructions. According to Carl Maria von Weber, Gottfried Weber's *Te deum* (1814) was the first fully scored work to employ the 'admirable method' of chronometrically marked tempos. To this claim Weber added that 'it is very much to be hoped that other composers will follow Herr Weber's lead in this matter'.[50]

As composers began to provide full scores of their works as a matter of course, they began, also, to specify with new precision the instruments upon which their works were to be performed. Rather than indicating that given notes be played on any instrument with a certain pitch range, composers began to specify the precise kind of instrument. Rather than composing music for particular, actually existing instruments and players, composers began to write music for instruments and performers at a distance. When composers began to produce music at a distance, they began also to consider musical instruments in abstraction, as separable from place and functional context. All these developments made it possible to assess the worth and contribution of an instrument to a given work of music on its own ideal and independent terms.

Thus, in the early nineteenth century, composers like Czerny and Berlioz produced meticulously worked out manuals on instrumentation, written from a standpoint at which they were able to consider the complexities of instrumentation in abstraction from the demands and

---

[49] *Roger North on Music*, 106. Ibid. for comments on early metronome use.
[50] Weber, *Writings on Music*, 128.

practical difficulties of actual performance. The same sort of distance was revealed also in the new treatises on orchestration. According to Elliott Galkin, Luigi Cherubini was the first composer 'to indicate exact and dissimilar dynamic levels for each instrument or group of instruments in an ensemble—in other words to consider balance of sound in the orchestra precisely in terms of its individual instrumental components'.[51]

VII

Little sense could have been made of the changes either in notational or instrumental precision, particularly those that moved notation away from being a more or less detailed map to being a full and complete representation of a work, had these changes been isolated phenomena. But they were not. They depended on other changes, some small, some large.

For example, when composers demanded around 1800 that they be given the power of dedication, they showed their unwillingness to be controlled in any way at all. Marking the transitional period, Haydn expressed his regret to Artaria and Company regarding 'one thing', that he did not have the honour of dedicating his Sonatas to the Demoiselles von Auenbrugger himself. Many of Haydn's contemporaries would not yet have found the courage to express their regret in public. And even Haydn went only so far. To the same company seven years after the previous letter was written, Haydn showed a traditional sort of respect for his employer.

I can think of no better and more fitting way to show my thankfulness to his Majesty, than by dedicating these 6 Quartets to him; but you won't be satisfied with that, because you will want to dedicate the[m] yourself, and to someone else. But to make amends for this loss, I promise to give you other pieces free of charge.[52]

Note the extent of Haydn's power. He had the right to sell his music at a price. But it was only when composers really began to feel that they owned their works, that they were able to demand unequivocally that they be given control over such small matters as dedication.

---

[51] *History of Orchestral Conducting*, 80.
[52] Letters of 20 Mar. 1780 and 19 May 1787, Haydn, *The Collected Correspondence*, 26 and 63.

And the same applies to the way they took control of how their works were titled. Since 1800, works have been titled in such a way as to indicate their status as independent, self-sufficient works. Thus, some works have been given titles indicating their status as completed, individuated works, inextricably connected to their composers, and devoted to purely musical matters. Beethoven's Symphony No. 5 in C minor, Opus 67 is an exemplary title. Other works have been titled in such a way as to reveal the composer's intent to produce music perhaps 'for the people', to show music's enormous range of expressive capabilities, or to emphasize music's ability to represent or express, even if not imitate, extra-musical phenomena. These titles have been of a 'programmatic' sort. But they have been used in the same way as titles of the absolute sort—ultimately to endorse the claim that music is a romantic and fine art.[53]

Titles for compositions had generally been of a different sort in earlier centuries. Usually they had reflected or mentioned the function for which the music was written, the personage to whom the composition was dedicated, or a place associated with the patron or performance. A Mass entitled *À l'usage ordinaire des paroisses pour les fêtes solenelles*, or music composed 'pour les soupers du Roi' are typical. Such titles continued to be used after 1800, but with less regularity in mainstream musical production.

Changes in dedication and titling procedures played a relatively minor role in the emergence of the work-concept, however. The changes that reinforced the distinction between composers and performers were more important. With this distinction clearly demarcated, musicians could in turn separate a work from its performance. In order for a work to acquire autonomous status, or, as Attali puts it, 'in order for music to become institutionalized as a commodity, for it to acquire . . . monetary value . . . it was necessary to establish a distinction between the value of the work and the value of its representation, the value of the program and that of its usage'.[54]

---

[53] Sometimes titles were given only after the composition of the work was completed. Despite Mahler's final judgement that his Third Symphony was absolute music, he provided, during the course of its composition, eight different programmes. He wavered constantly and without resolution over his use of titles, at one time replacing the programmatic extra-musical titles with musical titles. He knew that programmatic titles facilitated the public's understanding of his works. But he also felt that they could mislead the public, since they necessarily failed to capture music's essence being, as they were, constituted by words.

[54] *Noise*, 51.

Developments in copyright laws and publication helped 'institutionalize' works as commodities separable from their performances. Developments in notation helped free composers from involvement in performance. But these developments only partially account for the extent to which many composers became strategically uninterested in the performance of their works. A deeper source of that phenomenon takes us back to the romantic characterization of the emancipated composer.

Consider, first, how far some composers of this century have been willing to push the separability of work from performance. Early in this century, Arnold Schoenberg offered a bitter and defensive, though decisive, confirmation of such separability:

> I hold the view that a work doesn't have to live, i.e., to be performed at all costs . . . if it means losing parts of it that may even be ugly or faulty but which it was born with.[55]

Composers had supposedly obtained the right to respect for their decisions as to how their works should be performed. That their decisions, notwithstanding, were not always respected, forced them to escape into a world in which their works would no longer be touched by uncaring performers and interfering business managers. In 1918, to prevent the general 'cutting conspiracy'—the external tampering with or editing of his music by publishers and concert managers—Schoenberg founded the Society for the Private Performance of Music, a society devoted to composers and their works. The society closed after three years due to a lack of funds. Still, Schoenberg's solution was no different, as we shall shortly see, from Weber's solution adopted a century earlier.

Charles Ives also upheld the independence of his works from their performances. Influenced by American transcendentalists, he argued that music is a transcendent language whose concrete, mundane instantiation is little more than an offence. 'Why can't music go out in the same way it comes into a man,' he wrote, 'without having to crawl over a fence of sounds, thoraxes, catguts, wire, wood and brass?'[56] And for a different sort of evidence of the separability of work from

---

[55] Letter to Alexander Zemlinsky, 20 Mar. 1918, *Letters*, ed. E. Stein (London 1964), 54.
[56] *Essays before a Sonata and Other Writings* (New York, 1961), 8 and 84. See Dahlhaus's reference to the ' "invisible orchestra" concealing the mundane origins of transcendental music' (*Nineteenth-Century Music*, 394).

performance, consider Erik Satie's graduation diploma of 1905, from the Schola Cantorium, that stated that he had 'fulfilled the conditions required for devoting [himself] entirely to composition'.[57] How honoured and relieved Bach might have been had his livelihood not depended upon his acceptance that 'man must live among imperfecta', had he been able to exclude himself from the 'mundane' aspects of performance.[58]

There are statements to be found at the start of the nineteenth century as well regarding the devotion a composer could show towards the composition of works with novel disregard for their performances. About Beethoven it was written that:

> To show how little thought [he] gave to those who were to execute his music, we only need examine the *grande sonate* for piano and violin dedicated to his friend Kreutzer. This dedication might also be taken for an epigram, for Kreutzer played all his passages *legato*, and always kept his bow on the string; now, this piece is all in *staccato* and *sautille*—and so Kreutzer never played it.[59]

Even if Beethoven expressed disregard for particular performers rather than for performances, the point indirectly stands. The same point is made in another way by noting that composing unperformable or unplayable music was a way of showing disregard for a work's performances. Unplayability was a notion that arose just when the composition of works was sharply enough differentiated from the production of performances for the idea of writing notes independently of their performance to be able to feature in compositional choices.

This last point does not suggest that early musicians composed music that performers had no difficulty in playing. That would be untrue. Nor do I want to deny that modern composers took and still usually take great care to write for performers and performance. The point is only to show how, at one time, the idea of unplayability would have made little sense to those who were composing directly for the performance, for those, furthermore, who were normally present and in charge of the performance. The idea of unplayability would have

[57] P. D. Templier, *Erik Satie*, tr. E. L. French and D. S. French (Cambridge, Mass., 1969), 27–8.

[58] Following the brawl with the bassoonist, Bach was told 'man must live among *imperfecta*, that he must get along with the students, and that they must not make each other's lives miserable' (David and Mendel, (eds.), *The Bach Reader*, 51).

[59] 'The Baron de Trémont' (1809), Sonneck (ed.), *Beethoven*, 75. There is evidence also that Beethoven admired Kreutzer.

had force only for composers who could be sufficiently separated from, and, if they desired to be, unconcerned about the actual performance of their works. 'I am delighted to add another unplayable work to the repertoire,' Schoenberg once announced. 'I want the concerto to be difficult and I want the little finger to become longer. I can wait.'[60]

On the other hand, composers of works could not become completely uninterested in performances. For the emerging conception of their compositional activity depended heavily upon a suitable and compatible contrast emerging on the side of performance. Thus, given aesthetic attitudes of the time, musical works as abstract constructs *required* adequate realization in performance if they were to prove themselves worthy of being called 'works of fine art'. Adequate realization depended upon there being interpreters of works devoted to the task of realizing works through the medium of performance. The ideal of *Werktreue* emerged to capture the new relation between work and performance as well as that between performer and composer. Performances and their performers were respectively subservient to works and their composers.

The relation was mediated by the presence of complete and adequate notation. The only duty of liberated composers was to make it possible for performers to fulfil their role; they had a responsibility to make their works performable, and they did this by providing complete scores. This duty corresponded to the need, captured in aesthetic theory, for reconciliation of the abstract (the works) with the concrete (the performances). The comparable duty of performers was to show allegiance to the works of the composers. To certify that their performances be of specific works, they had to comply as perfectly as possible with the scores composers provided. Thus, the effective synonymity in the musical world of *Werktreue* and *Texttreue*: to be true to a work is to be true to its score.

Though performance was never seen to be just a matter of playing the notes right, even the additional demand for interpretation was

---

[60] Schoenberg's comment made in reference to his Violin Concerto (J. Machlis, *Introduction to Contemporary Music* (New York, 1961), 352). The idea of unplayability was expressed in Berlioz's bitter note written on the MS of *The Death of Orpheus*: 'This work declared "unplayable" by the Music Section of the Institute—Played on July 22, 1828.' Part of Berlioz's intention was to suggest that the declaration was unfounded; after all, the work did receive a performance. Another was to express his resentment at having lost a competition for which the work had been entered. For more on this, see Barzun, *Berlioz and his Century*, 63–4.

severely constrained by the ideal of fidelity to the work. Performers should interpret works in order to present the work as it truly is with regard to both its structural and expressive aspects. Room was to be left for multiple interpretations, but not so much room that interpretation would or could ever be freed of its obligation to disclose the real meaning of the work. A performance met the *Werktreue* ideal most satisfactorily, it was finally decided, when it achieved complete transparency. For transparency allowed the work to 'shine' through and be heard in and for itself. Such was the idea behind George Bernard Shaw's remark of 1888 that the pianist Sir Charles Hallé is always assured of an audience because he gives 'as little as possible of Hallé and as much as possible of Beethoven. When Beethoven', by contrast, 'is made a mere *cheval de bataille* for a Rubenstein', Shaw concluded, 'the interest is more volatile.'[61]

Even if Shaw were to have expressed a preference for Rubenstein's performance, it would still have been true that he chose to articulate the difference between the performers in terms of their being more or less true to the work. And such an approach concurs with the sort of stipulations made earlier in the century by, for example, Hoffmann. He had decreed, as I mentioned in my Introduction, that performers should 'live only for the work' and not make their 'personalities count in any way'.[62]

VIII

Performing under the ideal of *Werktreue* generated its own tension. Just as derivative forms of composition had emerged to satisfy the desire to use pre-existing musical materials, an alternative concept of performance emerged to satisfy the performer's need to perform without restrictions imposed by composers. Liszt, for example, spent a considerable part of his musical life developing two distinct forms of performance, first, performances committed to faithful renditions of

---

[61] *London Music in 1888–9 as Heard by Corno di Bassetto (Later Known as Bernard Shaw) with Some Further Autobiographical Particulars* (London, 1937), 41.

[62] Notice how Hoffmann combines the ideals of transparency and untouchability in order to compare early and modern performance practice. He writes: 'Like all compositions conceived by a master, the work demands to be received by the whole soul . . . with the precise knowledge of the transcendental implied by the melodies. That there would be certain embellishments . . . was [once] clearly understood. . . . But is it not wonderful that a tradition has been established concerning the precise embellishments of the song . . . so that today no one would dare introduce a foreign embellishment without censure.' (Schafer, *E. T. A. Hoffmann and Music*, 56.)

works, second, virtuoso performances devoted to the art of extemporization and the show of impressive performance technique.

Regarding the former, Berlioz once remarked on how Liszt performed in such a way that 'not a note was left out, not one added . . . no inflection was effaced, no change of tempo permitted'. In making 'comprehensible a work not yet comprehended,' Berlioz concluded, Liszt proved himself a 'pianist of the future'.[63] But, given the restraints of the *Werktreue* ideal, the only opportunity performers had to show their performing talents lay within the limited confines of the new cadenzas. For many performers—and Liszt was one of them—it seemed that this brief moment of 'show' was too brief and that premeditated performance was unduly restrictive. So they developed a practice to contrast with that of performing pre-composed, pre-meditated, or pre-concerted works.

The practice was based on the idea that performers could produce 'free and spontaneous' extemporized performances. If performers could not extemporize, they would compose or commission works designed just for virtuoso performance. Such free and spontaneous performance was defended on the grounds that it was inspirational and gave musicians immediate access to the world of transcendent truth.

In the early nineteenth century, however, attitudes towards extemporization and virtuosity were rather different from what they had been earlier on. Extemporization was not now generally thought to approximate to the condition of composition 'proper' and less respect was gradually given to the virtuoso performer, who had quickly come to be associated more, as it was said, with 'charlatanism' than with 'the legitimate objects of art'. Reminiscent of ancient attitudes, virtuoso performance had fast become regarded as a popular spectacle comparable to a circus act put on for those who preferred to 'hum and ha and applaud and feel astonishment at the feats of an instrumentalist'. Liszt did much, though not perhaps all he might have done, to temper a growing inclination on many a musician's part to overpopularize the performer's art. He did much to avoid the accusation that his art had come to serve 'only the self-serving virtuoso'.[64] Niccolo Paganini, whatever his intentions, had already done much to support the opposite view, if not just with his remarkable 'wizardry' on the violin, then also

---

[63] Walker, *Franz Liszt*, 236.
[64] Ibid. 168 ff. Also Raynor, *Music and Society since 1815*, ch. 4.

with his 'devilish' appearance.[65] By 1814, Tomaschek had already complained about

> the pianists of to-day, who only run up and down the keyboard with passages they have learned by heart. . . . What does that mean? Nothing! The real piano virtuosos [of old], when they played, gave us something interconnected, a whole. When it was written out it could at once be accepted as a well-composed work.[66]

Apparently modern piano performances, and most probably the extemporized performances amongst them, did not merit even being called 'badly composed works'.

Understanding of extemporization had altered over the centuries largely because of changes in notational practices. Before notation was used with any impact, the practice of music almost entirely consisted of musical extemporization; it almost was just the simultaneous composition-performance of music. That is to say, excluding repetition based on memorized melodies, performers generally extemporized. When notation was introduced to give concrete form to the idea of preserving written music, extemporization was limited to connote spontaneous composition-performance of, say, 'divisions on a ground', variations over a repeated figured bass. Over the years it became increasingly limited. By 1800, when composition was defined as involving the predetermination of as many structural elements as possible, the notion of extemporization acquired its modern understanding. For the first time it was seen to stand in strict opposition to composition 'proper'. In these terms, it is possible to see how changes in notation not only curtailed the liberties of the performer, but in close connection became the guard against extemporization. In fact, it has been said in retrospect that 'the whole history of composed music from John Dunstable to Beethoven may be described as the process of making the composer's defences sure against the incursions of the extemporizer'.[67]

---

[65] Cf. Louis Spohr's comments on both the negative and positive publicity surrounding Paganini's 'wizardry' and 'charlatanism', in his *Autobiography*, tr. anon. (London, 1865), 279–81.

[66] 'Johann Wenzel Tomaschek' (1814), in Sonneck (ed.), *Beethoven*, 105.

[67] H. C. Colles, 'Extemporization or Improvisation', in Grove, *Dictionary*, 5th edn., ii. 991.

IX

Conducting was also an activity that changed in the nineteenth century to accommodate the demand for faithful performances of works. The rise of modern conducting was marked not only by the standardization of time-beating techniques and the rise of the silent time-beater, but also by new demands to interpret and express musical works through performance. Concurrent with the emancipation of instrumental music, conducting became increasingly defined in association with the new demands of symphony orchestras.

Thus, after 1800, conducting was no longer thought to be just a matter of marshalling the beat, but of leading the orchestra in such a way as to interpet, express, and convey the musically meaningful content of a work. Every beat, it was decreed, must also be an aesthetically significant gesture. This new aesthetic, as opposed to the older functional, conception incorporated a novel distinction between two aspects of conducting: the 'technical' and the 'expressive'. Berlioz accordingly identified 'precision, flexibility, passion, sensitiveness and coolness combined, together with an indefinable subtle instinct' as the indispensable qualities of the new conductor.[68]

The rise of conductors was part and parcel of an all-embracing change. The emancipation of instrumental music increased the importance of rhythm and dynamics as independent musical elements. Formerly a function of syllabic and dramatic structure, these elements were newly determined according to purely musical criteria. This founded a new responsibility for temporal and dynamic control of orchestras. The same emancipation also served to found the emergence of the concert orchestra, later to become known as the symphony orchestra, and this helped give substance to an emerging distinction between orchestral and choral or operatic conducting. Surely, it was suggested, as a defence for what was a new profession, the task of presenting and conveying the musical content of a purely instrumental work must differ from the task of keeping the various elements of a choral or operatic production together.

As orchestras grew in size their leaders moved away from their seats at the harpsichord to take their place on the podium. As instrumental players received instruction on their role and position on stage, so conductors learned of their obligation to turn around to face the

---

[68] Galkin, *History of Orchestral Conducting*, 551.

players rather than the audience. The demands of the new music, the new grand-scale symphonies, required a very new sort of attention and leadership. Orchestra players needed more and more guidance in reading scores.

One might think that composers would have helped, but increasingly they did not. Their isolation from the performance of their music had rigidified distinctions not only between themselves and performers but also between themselves and conductors. The French critic, Castil-Blazé, stated in 1821 something we take for granted today, that the conductor 'must know how to compose even though he does not always compose the music which he plays'.[69]

Like the position of the modern performer, that of the conductor has not been an easy one. Subservience to composers and their works has generated pressures both political and aesthetic. Unlike performers, however, conductors have not developed an alternative art all of their very own. Conducting extempore is definitely not an option, even if 'virtuoso' conducting is. Still, they have managed over time to find a position within the world of classical music that is or should be much to their liking. For that story we have, however, to await the next chapter.

X

Just as transparency through fidelity was the ideal that regulated performing and conducting, the same ideal was decreed to regulate audience behaviour. Like performers and conductors, audiences were asked to be literally and metaphorically silent, so that the truth or beauty of the work could be heard in itself. But such attention was possible only if music was performed in the appropriate physical setting. For how could one listen attentively and in silence if there were distracting elements all around? Performances had not only to become foreground affairs, but they also had to be cut off completely from all extra-musical activites. It was with these sorts of ideas in mind that concert halls started to be erected as monuments and establishments devoted to the performance of musical works.[70]

---

[69] Galkin, *History of Orchestral Conducting*, 202.

[70] Most concert orchestras operated within operatic establishments. See Galkin, *History of Orchestral Conducting*, 58, and A. Carse, *The History of Orchestration* (New York, 1964), 168. By 1844, Galkin points out, there were still only four concert halls functioning independently of operatic and dramatic establishments. The best known of these was the home of the *Gewandhaus Orchester* in Leipzig, which since 1781 had had

In these buildings, as well as in the private 'museums' or societies, audiences began to learn how to listen not just to music but to each musical work for its own sake. A given performance of a work ceased to be interrupted by a long interval between movements, and audiences gradually ceased to participate in the way they had earlier on. The general desire for a quieter, more considerate, and more attentive audience was part and parcel of the growing respect for a new and 'civilized' musical event.

In 1739, Richard Wesley expressed an early understanding of the desire for silent reception when he proposed that

> in Publick Entertainments every one should come with a reasonable Desire of being entertain'd themselves, or with the polite Resolution, no ways to interrupt the Entertainment of others. And that to have a truce with Dissipation, and noisy Discourse, and to forbear that silly Affectation of beating time aloud on such an Occasion is . . . a great Compliment paid to the divine Author. . . . I think a profound Silence [is] a much more proper Expression of Approbation to Musick, and to deep Distress in Tragedy, than all the noisy Applause so much in Vogue, however great the Authority of Custom may be for it.[71]

Mozart would have approved. He would have approved even more had he seen in his lifetime practical evidence of Georg Gottfried Gervinus's statement of 1868, that 'every instrumental work has every claim on every kind of respect, provided that it is addressed exclusively to the only audience deserving of it, namely the *cognoscenti*—those who know and are in a position to judge its formal worth'.[72]

It took many years after the initial building of concert halls before audiences learned how to be quiet and how to listen, assuming that they (generally speaking) ever learned how at all. In 1809, Samuel Wesley remarked on the tension experienced by the new audience in a letter to one Mr Jacobs:

> You will wish to know how the Performances were received; and I wish you had been among us to have witnessed the Delight they afforded to the whole Audience, who (when at the Church) seemed to long for the Privilege of clapping and rattling their Sticks; even as it was, there was a constant Hum of Applause at the conclusion of every Piece, and there never could have been

the honour of being the first and only establishment devoted *exclusively* to the performance of concert music.

[71] Hogwood, *Handel*, 158.
[72] Bujić (ed.), *Music in European Thought*, 91–2.

more strict and flattering Attention any where than was manifest throughout the whole.[73]

Not all composers were so charitable. Expressing his distress over what was increasingly considered an 'impolite' audience, if not a downright rude one, Schumann wrote, in the voice of Florestan:

I've got you all together again, dear public, and can set you at each other's throats. For years I've dreamed of organizing concerts for the deaf and dumb, that you might learn to behave yourselves at concerts, especially when they are very beautiful. You should be turned to stone pagodas, like Tsing-Sing.[74]

If listeners were being trained to listen to music as best they could, critics were being trained to have even more developed an ear. For such was needed if they were properly to evaluate a work in isolation from any particular performance. Thus, it was suggested, the most ideal listeners—who would probably also be the critics—must move, like good performers, beyond their own personal interests and beyond the particularities of a performance to reach the work itself. In a review of Hoffmann's *Urdine*, Weber made the same point:

In order to judge properly any work of art that depends on a performance within the temporal dimension, it is essential for the critic to have a tranquil and unprejudiced mind, open to every kind of impression but scrupulously free of definite opinion or emotional inclination, except of course a conscious preparedness to accept the subject matter in question. It is only in this way that we can give the artist complete sway over our state of mind and enable his emotions and characters to transport us into the world of his creation.[75]

In earlier times, musical criticism of the sort to be found in Mattheson's and Mizler's writings had focused on theoretical specula-tions, compositional styles, and the effect of a given performance. Remarks on the inherent beauty and value of music were increasingly made, but for a long while they remained subordinated to extra-musical considerations. In 1753, however, William Hayes suggested that if we desired to find Handel's true character we should look to his compositions.[76] This was almost a radical statement in its time. As the

---

[73] *The Wesley Bach Letters*, 34.

[74] *Schumann on Music*, 65. Horowitz, in his *Understanding Toscanini*, notes that in 1895 at La Scala, Toscanini made women take off their hats, darkened the house, and outlawed encores (p. 46); and that Mahler would glare at latecomers and whisperers (p. 47). For discussion of comparable audience behaviour in North American concert halls, see L. W. Levine, *Highbrow/Lowbrow: The Emergence of Cultural Hierarchy in America* (Cambridge, Mass., 1988), 104 ff.

[75] Weber, *Writings on Music*, 201. 　　　　　　[76] Hogwood, *Handel*, 276.

century drew to a close this sort of statement had already taken on the appearance of obviousness. Indeed, what else should one look to?

The new form of music criticism, and what came at this time to be called analysis, took one step further than Hayes. It was possible to assess not just a composer's entire musical output but also any one of his individual compositions in isolation from any other. Hoffmann's highly romantic and Hanslick's formalist essays became exemplary of modern forms of music criticism. The influx in addition of companion treatises, whose sole concern was the articulation of notions of internal unity, form, and content, provided critics with a new language in which to express their appreciation or disapproval of each particular work. Thus, when Weber encouraged critics to 'follow a score and tak[e] the liveliest interest in the correctness of performance' in order to reach a fair evaluation of the work, he was assuming that the rationale behind this activity rested upon the belief that the work was something that merited attention as a distinct product of compositional activity.[77]

Finally, like every other activity at the time, music was affected in new ways by the unfavourable elements of its worldly or social practice. Weber warned his readers of the dangers of being a critic in an imperfect and messy world.

The judgement of works of art is often one-sided and dictated by party spirit, and publishers even hire writers to praise their publications, so that it is often difficult to obtain a proper hearing and evaluation for works of genuine worth by those who have not yet made a name for themselves.[78]

What was Weber's solution? It was the same as that taken later by Schoenberg: to form a private society for 'mutual support and action in the cause of art', with a rarefied atmosphere separated from the pressures of the mundane.[79]

XI

The erection of concert halls helped change expectations as to how an audience should behave. It also influenced a change in the style and procedure of concerts. In the latter, Liszt was one of the most important players. He is said to have introduced the term 'recital' in an 1840 concert in London at the Hanover Square Rooms.[80] Recitals,

---

[77] Weber, *Writings on Music*, 66.   [78] Ibid. 60.   [79] Ibid.
[80] According to Henry Raynor, the term was also used in Dunneley's *Dictionary of Music* in 1825 to refer to a solo vocal performance (*Music and Society*, 62).

previously involving the recitation of words, now became purely musical affairs.

Liszt was also largely responsible for putting a performer's name (independently of the composer's name) on a concert programme. Formerly musicians or 'concert givers' had not usually been mentioned by name unless they were also the composers, in which case their name might have appeared somewhere in small print. Liszt was also the first to play a whole programme of pre-composed works from memory, which indicated many things, not the least significant being that he had obviously practised the works before the concert took place.[81]

An obvious difference between old and new programmes (written or played) was the degree of respect accorded a complete instrumental work. One way to show appropriate respect was not to interrupt a performance of a symphony with too long a pause, interval, or intermission. The symphony should be played complete and at one sitting. Thus, in the preface to his 'Scottish' Symphony, Mendelssohn explicitly expressed the desire that the work should be played without long pauses between the movements, in the interest of its unity.

The length of concerts changed accordingly. As audiences began to listen to the music without distractions, programmes were reduced in length to accommodate listeners' attention spans. Formerly, the amount of music performed had been decided by the length of the extra-musical ritual. Composers had simply composed and performed enough music for the occasion. That had resulted in the performance of a greater quantity of music within a single 'concert' than has commonly been heard in concert halls in the last 200 years.

Written programme notes or ('programme books') for concerts also began to acquire their modern and standardized form. Works, composers, and performers were now the main items specified, and the accompanying notes came to be designed to help an audience adopt a suitable aesthetic attitude. If the music was being called programmatic, the notes would help the audience, as Liszt said, to grasp the 'correct poetical idea'. If the music was being called absolute, notes could merely be of an analytical or biographical sort. Of course, in a certain sense, no verbal notes of any kind were necessary for adequate reception of music, whether absolute or programmatic. What audiences really had to do was listen.[82]

---

[81] Walker, *Franz Liszt*, 285 and 80.

[82] Cf. Berlioz's instructions for programmes to be distributed at performances of his *Symphonie fantastique* (1830), (Leipzig, 1900), 1.

XII

Erecting concert halls and forming private societies were not the only ways musicians sought to institutionalize the new ideals of a work-based practice. These ideals were also to be empowered within the academy. Thus the new interest in music, autonomously conceived and severed from its connections with extra-musical institutions, was marked by the founding throughout Europe of both musical academies and public societies. In Britain alone, the Philharmonic Society was formed in 1813, the Royal Academy of Music in 1826, and the Society of British Musicians in 1834, and these by no means form an exhaustive list.

Biographies of composers also began to be produced, beginning with the lesser-known *La Vie et les ouvrages de Piccinni* by Ginguené in 1800. Forkel's biography of Bach followed in 1802, and this was followed shortly afterwards by biographies of Haydn by Carpani in 1812, and of Mozart by Nissen and of Palestrina by Baini both in 1828. It had not been common to find a biography written about a single composer in earlier times. The notable exception, perhaps, is the one suggested by Peter Kivy to be the first modern biography of a composer, Mainwaring's *Memoirs of the Life of the Late George Frederic Handel* (1760).[83] A less notable exception—if one can even call it an exception—is Walther's *Musicalisches Lexicon* (1732), which contains a few biographical notes about composers.[84]

A new sort of music history began to be written as well, exemplified in Abraham Rees's *General History of Music* of 1798 and his *Cyclopedia* of 1802. For the first time music history concentrated on 'great names' and 'masterpieces', thereby replacing Preatorius's seventeenth-century history *Syntagma*, and J. Bonnet's *Histoire générale de la musique* of 1715/25, which offered histories of musical functions, uses, and styles. There was, finally, a marked increase in the production of bibliographies and music journals, the most significant of which were Forkel's *Allgemeine Literatur der Musik* of 1792, Friedrich Rochlitz's *Allgemeine musikalische Zeitung* of 1798, and Robert Schumann's *Neue Zeitschrift für Musik* of 1834. In England the *Harmonicum* appeared in

---

[83] 'Mainwaring's Handel: Its Relation to English Aesthetics', *Journal of the American Musicological Society*, 17 (1964), 170.

[84] David and Mendel (eds.), *The Bach Reader*, 26. On several accounts, Burney's *General History of Music* written in 1789 is transitory in style. For further, related details, see A. Hyatt, 'General Musical Conditions', in Abraham (ed.), *The New Oxford History of Music* (10 vols.; Oxford, 1986), viii. 1–25, and Allen, *Philosophies of Music History*, ch. 6.

1823, soon followed by the *Musical Library*, and *Musical World*. All these were devoted, if not exclusively then more so than ever before, to the mainstream production of large-scale instrumental works, as opposed to choral and operatic music, so dominant in earlier times.

XIII

The ideal of *Werktreue* pervaded every aspect of practice in and after 1800 with full regulative force. Following from the central conception of a musical work as a self-sufficiently formed unity, expressive in its synthesized form and content of a genius's idea, was the general submission of all associated concepts. Concepts and ideals having to do with notation, performance, and reception acquired their meaning as concepts subsidiary to that of a work. In a certain sense this had had to be the case, for these subsidiary concepts had served to give a highly abstract concept concrete expression. Without their development, in other words, the abstract work-concept developed and articulated within the romantic, aesthetic theory of fine art would never have found its regulative force in practice.

The history of music in the late nineteenth and twentieth centuries does not prompt the simple conclusion, however, that further developments merely added to the force of the work-concept. For as fast as the work-concept found its force, so oppositional concepts and theories began to emerge to challenge it. As oppositional concepts and theories emerged, the work-concept in turn began to exert its full force. The consequent battles and tension between the work-concept and its opposition is the major theme of my final chapter.

9

Werktreue: Confirmation and Challenge in Contemporary Movements

This chapter weaves remaining threads together, but it does not reach a final conclusion. It is designed to leave readers sharing its author's sense of wonder at how human practices come to be, succeed in being, and continue to be regulated by one set of ideals rather than another. Wonder can increase rather than be diminished despite a philosophical and historical understanding of these ideals. Often such understanding accentuates while it explains the peculiarity of human practices; hence, the limits of explanation. In the end I hope to leave readers with the specific feeling that speaking about music in terms of works is neither an obvious nor a necessary mode of speech, despite the lack of ability we presently seem to have to speak about music in any other way.

In the last two centuries, there have been many debates over the force of the *Werktreue* ideal (otherwise referred to as the ideal of fidelity or authenticity), and all of them point to tensions generated by the ideal. To be sure, the debates have not always resulted in direct confrontation between those who support and those who oppose the ideal or the related work-concept. Often, questions about *Werktreue* have been treated vicariously, as part of a debate about something else or, locally, within a debate over a particular area of musical practice. Still, somewhere and somehow the *Werktreue* ideal has been in question.

In this chapter, I shall investigate the terms of these debates. But to do only this would not leave us with as much philosophical understanding of the work-concept as I promised earlier. So, interspersed between historical remarks about *Werktreue*, I shall offer some final philosophical remarks about open concepts. That the latter remarks emerge directly out of the historical ones is neither just convenient nor coincidental. They are placed here to bring the methodological argument of the book full circle, to demonstrate, finally, some of the ways in which history and philosophy can become inextricably connected with one another.

I

Nowadays, no form of musical production is excluded a priori from being packaged in terms of works. Early music—the music of Palestrina, Vivaldi, and Bach—is packaged in this way, though the music in question was not so packaged at its moment of origin. There is a tendency, also, to classify most if not all 'experimental' music as works. We speak of the works of John Cage, Max Neuhaus, and Frederic Rzewski, even though these musicians do not think of themselves as composing within the romantic tradition. 'It's a very deliberate step of mine', Neuhaus writes, 'not to record the pieces. These pieces are not musical products; they're meant to be activities.'[1]

We often disregard the conceptual differences between a work and an improvisation or those between a work and a transcription. Both improvisation and transcription emerged with their modern understanding as concepts sibling to the work-concept. They stand in an intimate relation to the latter, but it is not one of identity. It seems to be indeterminate. On occasion we refer to transcriptions as transcriptions; sometimes we speak of them as works in their own right. Similar indeterminacy obtains for improvisations.

Those interested in 'found art' might speak of the phenomena of natural music—bird songs, for example—if not as works, then at least under the label of music. But suppose recordings were made, on the basis of which scores were produced, the presence of which then made it possible for us to perform the music on suitable instruments. How easy it would now be to talk of these phenomena as works.

When we speak of works we do not think immediately of jazz, folk, or popular music, nor of music accompanying other art-forms, such as film and dance music. Nor do we think immediately of music purposefully integrated into the everyday world, say, the music of religious services or of other rituals. But this does not exclude the possibility of thinking about these types of music in terms of works. We usually speak of film music in terms of musical or film *scores*. But we could call these scores *works*, if they were marketed and evaluated independently of the films (recall Prokofiev's *Alexander Nevsky*). We could think also of a performance based on an Indian raga as a performance of a work were it to take place, say, in the Royal Albert Hall as part of a celebration of 'the music of other cultures'.[2]

---

[1] 'Max Neuhaus', Rockwell, *All American Music*, 146.

[2] The history of film music provides a mini-version of the larger history of music. In

All in all, the contemporary use of the work-concept does not confine us to 'concert' music produced since 1800. We are tempted to understand music of many kinds as involving the production of works. But if the work-concept emerged as a result of a specific and complicated confluence of aesthetic, social, and historical conditions, *why* have we wanted to, and *how* have we been able to extend the employment of the work-concept seemingly so pervasively?

II

First, an answer—or at least a major part of one—to the 'why' in the question: the view of the musical world the romantic aesthetic originally provided has continued, since 1800, to be the dominant view. This view is so entrenched in contemporary thought that its constitutive concepts are taken for granted. We have before us in fact a clear case of *conceptual imperialism*.

It all began around 1800 when musicians began to reconstruct musical history to make it look as if musicians had always thought about their activities in modern terms. Even if it was not believed that early musicians had thought explicitly in these terms, the assumption was that they would have, had circumstances allowed them to do so. Reconstructing or rewriting the past was and remains one of the most characteristic ways for persons to legitimate their present, for the process aids in the general forgetfulness that things could be different from how they presently are.

'It is certain,' Forkel wrote in 1802, 'that if the art is to remain an art and not to be degraded into a mere idle amusement, more use must be made of classical works than has been done for some time past.' When had it ever been done? Not that often. In 1768, a society had been founded in London to promote the performance of ancient music. In 1807, Weber spoke of the new Dresden Singakademie specializing in the performance of neglected classics. But the activities of these and

hindsight, it can be viewed as a struggle for musicians to gain independent recognition of themselves and their music. Composers producing music for Hollywood films often complained of music's subservient role, that it was accorded importance only in relation to the drama and text it accompanied. Many complaints mirrored those made by musicians in earlier centuries who sought to have music received and judged on its own terms, or at least as an equal partner in a multi-art production. In the history of film, additional tension was caused when composers used pre-existing works which then had to be adapted (and in composer's minds, usually ruined) to fit the film. For further discussion, see R. M. Prendergast's *Film Music: A Neglected Art* (New York, 1977), esp. 70 ff.

other similar societies did little to alter, in the short run at least, a mainstream attitude of the time that music of the present was the only music worth listening to. Joseph Horowitz recently explained that, even in the 1840s, 'most Viennese and Parisian concertgoers denigrated the notion that the greatest music might be music of the past', and that it was only in the 1850s that the public's attitude started to shift. If this was true of the general public, however, it was not true of prominent musicians, such as Liszt, who, already in the 1830s, was performing the music of past masters. According to Alan Walker, Liszt started a trend for pianists to include 'historical pieces' in concert programmes.[3] What is also clear is that the musician's interest in the revival of past music was taking a specifically romantic form.

Reconstructing the past was partly motivated by a new sort of academic interest in music history. Bringing music of the past into the present confirmed at least one tenet central to romanticism, that of replacing a traditional, static conception of nature with a dynamic conception of history. But there was another interest in reconstruction that was more influential. Musicians did not look back to the past, as they once had done, to find models for contemporaries to imitate. Instead, they began to see musical masterpieces as transcending temporal and spatial barriers. One level of history was being transcended to reach another. Works were not to be thought about as expressive or representative of concrete historical moments, but as valuable in their own right as transcending all considerations other than those of an aesthetic/spiritual nature. 'Do not think that old music is outmoded,' Schumann wrote in 1834. 'Just as a beautiful true word can never be outmoded, so a beautiful piece of true music'.[4]

One way to bring music of the past into the present, and then into the sphere of timelessness, was to strip it of its original, local, and extra-musical meanings. By severing all such connections, it was

---

[3] David and Mendel (eds.), *The Bach Reader*, 297; Weber, *Writings on Music*, 115; Horowitz, *Understanding Toscanini*, 135; Walker, *Franz Liszt*, 256. According to Horowitz, it was not until 1870 that orchestras consolidated a 'museum identity as showcases for dead masters'. The repertoire of the Gewandhaus Orchestra shows *percentages* of works by dead composers: 13% between 1781–85; 23% between 1820–25; 39% between 1828–34; 48% between 1837–47; 61% between 1850–55; 76% between 1865–70. For further and comparable discussion, see W. Weber, 'Mass Culture and the Reshaping of European Musical taste, 1770–1870', *International Review of the Aesthetics and Sociology of Music*, 8 (1977), 5–22, and J. P. Burkholder, 'Museum Pieces: The Historicist Mainstream in Music of the Last Hundred Years', *Journal of Musicology*, 2 (1983), 115–34.

[4] Quoted in David and Mendel (eds.), *The Bach Reader*, 370.

possible to think of it now as functionless. All one had to do next was impose upon the music meanings appropriate for the new aesthetic. Many musicians proceeded, therefore, to conceive of past music in the romantic terms of works. The canonization of dead composers and the formation of a musical repertoire of transcendent masterpieces was the result both sought and achieved.

Reintroducing early music into the modern repertoire as 'timeless masterpieces' gave to early composers and their music what they had never had in their lifetimes—precise notations, multiple performances, and eternal fame. It also provided for the musical world a constant, standard, and significantly enlarged repertoire of musical masterpieces. Constancy and standardization, processes which provided musicians with exemplars and standards by which to guage their own activities, overtook the museum of musical works at a rapid speed. The repertoire of the Classics or 'War Horses' became so standardized, in fact, that it took on, as someone once said of Beethoven's Fifth Symphony, the familiarity of a breakfast cereal.[5]

But there is a twist in the story that makes the formation of the musical museum all the more interesting. In the early nineteenth century, the idea of classical music was formed in contradistinction to that of romantic music and was used to refer to a style of music predominant in the mid- to late-eighteenth century. In initial classifications, the classical period extended roughly from Bach to Mozart, though Beethoven had at least one foot in the camp as well. Gradually, for a variety of complex reasons, music of the classical period came to be viewed as the most perfect and valuable sort that had ever been produced. The period producing masterpieces of classical form had been a 'Golden Age'.

Contrary to classifications of style or genre, however, musicians began to view classical compositions in romantic terms, as the most perfect works of *absolute* music. Beethoven, as the archetype romantic composer, produced paradigmatic examples of absolute works, albeit that they were of classical style. His works were then used as exemplars by which to judge and read the classical music that preceded him. This process resulted in a paradoxical play of terminology. As Dahlhaus writes:

The idea of absolute music was formed on the basis of works that first had to be reconstrued before they could even suit this category, a category whose full

---

[5] For discussion of rereading the past, see ibid. 135 and 371, and for standardization (and breakfast cereal) see Horowitz, *Understanding Toscanini*, 402.

significance was, in turn, revealed to the nineteenth century only by these very works.[6]

Friedrich Blume reminds us that 'it is a widely held but erroneous idea that the Classic era was "the age of instrumental music;" that, in reality, was the Romantic era.'[7] The impact of these terminological twists was enormous, for it lay behind the gradual extension of the name 'classical music' to cover all music of a 'serious' nature. Though it was originally intended to signify music written just before the romantic era, the name began to embrace a much larger class of compositions, gradually the entire class of musical works. All 'serious' music would henceforth fall within the sphere of 'classical' musical works.

Thus, when Mendelssohn introduced Bach's *St Matthew Passion* into the modern repertoire he, to put the point crudely, took the music away from the church and put it into the concert hall. What he did (though this proved a complicated task) was to fill in and update the score in a way true to Bach's 'intentions', so that the work could henceforth be performed under modern conditions. But as Mendelssohn, and others like him, began to reconstruct music of the past, they were not yet aware of the extent to which reading modern intentions into the minds of past composers could undermine their being true to the music itself. They did not see that, if one can in fact be true to anything at all, then being true to early music is not necessarily the same as being true to a work. Hence, the rising phenomenon in the nineteenth century of 'playing Bach in the spirit of Brahms'.[8]

Loyal and noble intent aside, this lack of awareness not only pointed to nineteenth-century problems regarding the scope of use of the *Werktreue* ideal. It also points to a blatant manifestation of conceptual imperialism pervasive in more recent times. Thus, many persons, convinced nowadays by the greatness of classical music, have found reason to describe all the types of music in the world, of

---

[6] *Foundations of Music History*, tr. J. B. Robinson (Cambridge, 1983), 28.

[7] *Classic and Romantic Music*, 67. See also pp. 8–17 for further discussion of classic-romantic terminology and its connection with romanticism.

[8] Cf. H. Haskell, *The Early Music Revival: A History* (London, 1988), 13–16, and David and Mendel (eds.), *The Bach Reader*, 370 ff. And note, here, a typical case of a modern editor filling in the term 'work' to replace the unstated, 'vaguer' term 'Musick'. Speaking of activities that diverted himself, Roger North writes: 'I had my trinketts there to play with; as for instance, Musick, which was a very great enterteinement to me, not more by the practise, but [by] wrigting over into books such [works] as I could collect and thought good.' (*Roger North on Music*, p. xix.)

whatever sort, by means of a work-based interpretation. Such persons have believed that the closer any music embodies the conditions determined by the romantic work-aesthetic, the more civilized it is. For them, classical music is not only regarded as quintessentially civilized, but as the only kind of music that is. The same is believed of fine art more generally, so that it often serves as a standard when one wants to attribute positive value to something. Consider the pleasure a chef feels from being told that his Black Forest Gâteau is a 'work of art', or the satisfaction of a car manufacturer who produces a car deemed a 'fine artistic product'. The phrase 'musical work', like 'work of art', is used with evaluative as well as classificatory sense. What we see under imperialistic influence is a conflation or contamination of the two senses.[9]

The direct result of seeing the world's music—including early music—through romantic spectacles is that persons have assumed that the various types of music can easily be packaged in terms of works. Ways are sought to assign the 'works' to composers, to represent them in full score, thereby allowing them to be regarded as having a fixed structure with a sharply defined beginning and end, thereby enabling them to be performed on numerous occasions as part of a programme of works in the fine setting of a concert hall. Because this way of thinking leads to our alienating music from its various socio-cultural contexts and because most music in the world is not originally packaged in this way, do we not risk losing something significant when we so interpret it?

Do we not lose something when we hear the music of a flamenco or a blues guitarist in a concert hall? For the conventions of that setting determine that audiences should listen with disinterested respect to the 'works' being performed. The ideally silent audience cannot even tap its many feet—not without a certain discomfort at least.[10] Do we not lose something—even just an acoustic something—when we hear eighteenth-century chamber music performed, outside of the 'chamber', in symphonic concert halls? Critic Jonathan Keates, in thinking specifically about eighteenth-century operas, is most aware of the tendency for the romantic aesthetic to dominate.

[9] For an excellent discussion of the pervasive influence of the romantic aesthetic, see E. Panowsky, 'The Ideological Antecedents of the Rolls Royce Radiator', *Proceedings of the American Philosophical Society*, 107 (1963), 273–88.

[10] At a 1987 concert of flamenco guitar music played by Carlos Montoya, it was evident that the 'endings' of his 'works' were artificial. He simply stopped playing at a certain point and the audience applauded.

However much we might pride ourselves on our understanding of the eighteenth century, the values of the *Gesamtkunstwerk* continue to be applied with ingenious blindness, to appraising the overall worth of its art-forms. In their essentially fluid character, which ensured that no performance guaranteed a faithful replica of its predecessor and that structural alterations were not merely permissible but desirable and even expected, the operas resemble the most aleatoric of our contemporary modes.[11]

Conceptual imperialism has not been one-sided. *Werktreue* beliefs have increasingly been adopted by musicians involved in the production of many different kinds of music. Some jazz musicians, for example, have sought (perhaps only for financial reasons) and then found respect from 'serious' musicians by dispensing with the smoky and noisy atmosphere of the club and by performing instead in tails. Some, apparently willingly, have adopted the institutional conventions associated with 'serious' music. Many musicians in China have tried as well to reconcile their desire to compose and perform Western-style works with the demands of their own traditions—much to the annoyance often of the Chinese authorities. The composition in the 1970s, and subsequent debate over, the *Yellow River Concerto* is a case in point. A rather different example derives from Schoenberg's defence of his modern 'twelve-tone' method. He was once forced to remind his critics that composers who use his method compose just as Austro-German composers have always done. 'My works are twelve-note *compositions*,' he wrote to Rudolf Kolisch, 'not *twelve-note* compositions.'[12]

It is not surprising that musicians of many sorts have been forced or have felt a need to justify themselves to their critics by showing some willingness for their music to meet the conditions of work-production. It is hard to completely ignore the judgements of others. Think, in this context, of the effect the distinction between civilized and uncivilized

[11] Review of Reinhard Strohm's *Essays on Handel and Italian Opera*, *Times Literary Supplement* (31 Jan. 1986), 118. Keates's use of the term *Gesamtkunstwerk* is not antithetical to that of work termonology.

[12] On the *Yellow River Concerto*, see L. Rowell, *Thinking about Music: An Introduction to the Philosophy of Music* (Amherst, Mass., 1983), 218, and R. C. Kraus, *Pianos and Politics in China: Middle-Class Ambitions and the Struggle over Western Culture* (New York, 1989), 148–52; Schoenberg, letter of 27 July 1932, quoted in P. Griffiths, *A Concise History of Avant-garde Music: From Debussy to Boulez* (New York, 1978), 81. In the rest of the letter, Schoenberg contrasts the use of a 12-tone series with a single composition, to stress that the aesthetic qualities of the music reside in the composition and not in the series. He contrasts himself here to Joseph Hauer, 'to whom', he says, 'composition is only of secondary importance' (*Letters*, ed. Stein, 1964).

music, as well as that between artist and craftsman, must have on a musician's conception of his or her practice. Imagine how inferior Chinese musicians must feel on hearing the dismissive words of the romantic Berlioz:

The Chinese sing the way dogs bark, or the way cats vomit when they swallow a fish bone. The instruments they use to accompany the voice are resemblant of veritable instruments of torture.[13]

Or the jazz musicians who read in 1920 that '[j]azz is cynically the orchestra of brutes with nonopposable thumbs and still prehensile toes, in the forest of Voodoo.' And so the critic rambled on.[14] As recently as 1987, a public discussion was devoted to the claim that Duke Ellington 'utterly fails to conform to the criteria of the conventional idea of "the artist," just as his improvised productions fail to conform to the conventional view of the "work of art" ', even though Ellington saw himself 'as an "artist" in this sense and took to composing "works" for the concert hall'.[15]

Many musicians, moreover, whether they have wanted to criticize or defend a given sort of music, have used works produced around 1800, and usually by Beethoven, as the standard. This standard has generated all sorts of telling evaluations. Violinist Jascha Heifetz informed his public that he occasionally performed works by contemporary composers for two reasons: first, to discourage the composers from writing any more; second, to remind himself of how much he appreciated Beethoven.[16] In a recent *Newsweek*, an article entitled 'Bird Lives in his Incandescent Art' began with the judgement that 'Charlie Parker was the Beethoven of improvised music—a revolutionary as renowned for his personal furies as for his incandescent art.'[17] H. F. Chorley, critic for the *Athenaeum* wrote in 1856 after

[13] Quoted in A. Daniélou and J. Brunet, *The Situation of Music and Musicians in Countries of the Orient*, tr. J. Evarts (Florence, 1971), 21. Originally from Berlioz *A travers chant*, 'Mœurs musicale, de la Chine', 265: 'Il a une musique que nous trouvons abominable, atroce, il chante comme les chiens bâillent, comme les chats vomissent quand ils on avalé une arête; les instruments dont il se sert pur accompagner les voix nous semblent de véritables instruments de torture.'

[14] See Attali, *Noise*, 104.

[15] E. J. Hobsbawm, Review of *Duke Ellington* by J. L. Collier, *New York Review of Books* (19 Nov. 1987), 3–7. Think also of George Gershwin's 'classical compositions' in this context, and the discussion of them by L. Bernstein in his *The Joy of Music* (London, 1959), 59–61.

[16] *Life*, 1961, quoted in Crofton and Fraser (eds.), *A Dictionary of Musical Quotations*, 10.

[17] J. Miller, 31 Oct. 1988, 71.

hearing 'Madame Schumann's performance of Dr. Schumann's Concerto in A minor', that 'because we cannot fancy the Concerto adopted by any performer in London, we will forbear to speak of the composition as a work'.[18] Of a trio by Anton Webern, a critic once wondered almost in horror, 'what kind of organization a brain must have to be capable of such productions. One is reluctant to utter the word "abnormal," but on the other hand one cannot assert that there is any connection in this Trio . . . with our accustomed ideas about music.'[19] The standards held by these critics all come from somewhere, and usually from the same place.

Consider, finally, how critical some persons are of popular music, on the grounds that a given song has a simple form or that the music 'doesn't last', or that popular music is expressive of 'infantile emotions'. Yet a certain kind of simplicity is required in much of the world's music precisely because of its acknowledged social, political, or religious function. Such music is not designed to surpass time, or to stand its test.[20] It is used by particular social groups to express their socio-political desires. Its value and significance does not derive from a romantic aesthetic, nor, therefore, does its fair evaluation.[21]

The migration of concepts among different types of music is, of course, pretty much inevitable given our conceptual or cultural limitations. Persons find it easier to comprehend the music whose production they are not involved in by employing a familiar rather than an unfamiliar conceptual framework. Anyway, the assimilation of 'alien' concepts into a given type of music is not always bad. There is sometimes a healthy blurring of the boundaries between different types of music, a blurring that can be fostered by conceptual migration. Thus, even given, or despite, imperialistic influence, the adoption of

[18] London, 17 May, quoted in N. Slominsky, *Lexicon of Musical Invective: Critical Assaults on Composers since Beethoven's Time* (Seattle, 1965), 169.

[19] From the *Deutsche Allgemeine Zeitung* (Berlin, 24 May 1928), quoted in Slominsky, *Lexicon*, 214.

[20] For detailed discussion of this last phrase, see A. Savile, *The Test of Time: An Essay in Philosophical Aesthetics* (Oxford, 1982). For a concise introduction to alternative evaluative criteria used to assess non-opus, non-western types of music, see B. Nettl, *The Study of Ethnomusicology: Twenty-nine Issues and Concepts* (Urbana/Chicago, 1983), esp. chs. 6 and 8.

[21] Cf. S. Frith, 'Towards an Aesthetic of Popular Reception', in R. Leppert and S. McClary (eds.), *Music and Society: The Politics of Composition, Performance and Reception* (Cambridge, 1987), 133–49.

concepts into foreign types of music can lead—and has led—to new and interesting musical styles.[22]

Effects apart, the fact is that the work-concept with its conceptually dependent ideals of compliant performance, accurate notation, and silent reception has been adopted by many interpreters and producers of music of all sorts. Since 1800, not only have the most disparate arts 'aspired to the condition of music', but many disparate types of music have aspired to the condition of musical works.

III

So far I have described some of the reasons why the work-concept has come to be employed in settings seemingly set apart from its original context. Now I want to show how this has been possible.

The musical work-concept found its regulative function within a specific crystallization of ideas about the nature, purpose, and relationship between composers, scores, and performances. This crystallization shaped and continues to shape a standard or 'establishment' interpretation of the work-concept and of the practice it regulates. It continues also to motivate our classification of examples of musical works.

When we use the work-concept or any other open concept, we use it with an understanding revealed in our beliefs, ideals, assumptions, expectations, and actions. We can, however, use it in different ways. At least two of these ways ground a distinction between original and derivative examples. Original examples are not those produced first, but those produced directly and explicitly under the guidance of the relevant concept. We classify examples as derivative, by contrast, when

---

[22] For discussion of racial purity and impurity in musical styles and of the 'crossing and recrossing' of emigrating melodies, see B. Bartók, 'Race Purity in Music', *Modern Music*, 19 (1942), 153–5. One of the advantages of allowing different sorts of music to share concepts and values is that it makes comparisons between them possible. One wants to prevent concluding that a given type of music can only be assessed on its own internal terms, for then no judgement of it, as compared with another type of music, is possible. One wants to be able to look at different types of music, as one does different systems of morality, in comparative terms. This is possible only if one allows that a given type of music can be judged according to concepts and values that do not belong only to its particular form of production. Of course, concepts and values used in these comparisons do not need to carry imperialistic bias to function. Cf. a recent study of music education by Lucy Green (*Music on Deaf Ears: Musical Meaning, Ideology and Education* (Manchester, 1988)) that is largely devoted to revealing the continuing hegemony of classical music ideals in what is purportedly an education in all sorts of music.

we classify them as falling under a certain concept, even though these objects were not brought into existence with that concept in mind or within the specific part of practice associated with it. Whether an example is original or derivative is not something decided independently of how we *use* the concept. Thus, we should talk first of an open concept as having an original and derivative arena of employment and only then of its extension in terms of original and derivative examples.

The original use of an open concept tells a familiar story about conceptual use. Derivative use is more complicated, for there are different sorts of such use.[23] Given the musical work-concept, one may look first at how far the activities of musicians producing music of a non-classical sort have approximated none the less to the condition of producing musical works. One may look at how work-associated concepts have gradually been taken over and how musicians have then begun to speak of their production in terms of works.

When non-classical musicians borrow work-associated concepts (and only when they do, for they certainly do not always have an interest in doing this), they adopt an understanding sufficient to sustain the functioning of these concepts. In the romantic eyes of the classical musician, they more or less successfully impose the appropriate categories upon their practice. They act in what classical musicians consider to be the right way and in a way they themselves presumably find satisfactory and rewarding. The concepts come to be employed in a non-classical setting in much the same way as natives incorporate into their understanding any concepts introduced to them by foreigners. Without exposure to foreign concepts, native musicians probably remain oblivious to them. It is on this assumption that we say that the use of the concepts is foreign to the native's own practice, and that if the concepts are used at all, they are used derivatively.

Or one can look at how musicians steeped in the romantic view interpret non-classical types of music according to their own conceptual framework. This would be more like persons who, entering into a

---

[23] The distinction between original and derivative examples concerns the packaging of music. Unless one puts an imperialistic cover on the distinction, it is not to be equated with a distinction between legitimate and illegitimate use. Nothing in my use of the distinction speaks to the legitimacy of the use of a concept. Incidentally, in my article 'Being True to the Work', *Journal of Aesthetics and Art Criticism*, 47 (1989), 56–67, I used the term 'paradigm' in place of 'original'. I now think the latter term more accurately describes examples produced directly under the guidance of the work-concept. The term 'paradigm' has connotations of an example's being exemplary and clearly not all original examples are paradigmatic even if they are candidates for being so.

foreign cultural context, make use of their native linguistic or social apparatus to acquire what is for them a sufficient grasp of unfamiliar customs. Work-orientated musicians can effectively choose to regard music in terms of *works* if they believe they can, with the relevant understanding, act successfully or usefully in relation to the music regarded in this way. In general, they extend the concept's employment when they infer its presence in a cultural setting which, without that inference (or interference), would not acknowledge its presence.

Hence it is possible for musicians to look at a practice, one, for example, in which the music of Bach or Palestrina was produced, and to classify the music derivatively as works. This is possible because they can identify composers, represent the music in adequate notation, specify determinate sets of instrumental specifications, etc. In the same manner, it is possible for classical musicians to identify the same features when listening to transcriptions, improvisations, aleatoric pieces, or even blues songs, and to talk thereafter of the music in terms of works. In all these cases, the use of the work-concept is derivative.[24]

The assimilation of work-associated concepts into different kinds of music is not uniform. Some kinds of music stand in closer conceptual proximity to work-music than do others: the eighteenth-century sonata stands in closer conceptual proximity than the Indian raga. Some kinds of music are regulated by ideals that conflict with the *Werktreue* ideal. Jazz is an obvious example. It is likely, finally, that derivative works, precisely because they are derivative, fail to comply perfectly with the beliefs and expectations associated with the work-concept. Notations (or transcriptions), if existing at all, will not always be as precise as Beethoven would have liked. If they are, that often indicates that one has been rather lenient with the ideal of fidelity to the original music—the recent ethnomusicologist's complaint. When, however, we confront such imperfectly complying examples, we do not exclude them from falling under the concept. To do so would defeat the very idea of their being derivative.

---

[24] There is a fundamental indeterminacy or vagueness involved in this whole process which points sometimes to conflicts in aesthetical and musical values. Thus, one can ask: does the transcription of a work offer the public something independently valuable, such that it comes to be judged on its own terms independently of reference to the original work? Or should it always be judged in relation to the original work? The advantage of this sort of indeterminacy—the advantage of there not always being a clear answer to this question—should be obvious to those who find it valuable and interesting to judge a given composition both as an independent work and as a transcription.

When an example falls under a concept it is less a matter of its having the appropriate properties than of its being brought to fall under the concept by a user of that concept. If the relevant features are lacking in the first place, they can be assigned to the example so that it can be regarded in the right way. If this is not possible in any adequate manner at all, the attempt fails. Only then do we exclude the example from falling under the concept.[25] This procedure functions because of the *connection* obtaining between a concept's original and derivative uses. The connection is a conceptual dependency of the latter on the former, understood in terms of the aims and beliefs of musical agents, how and why an agent wants to use a concept in a given way.

Thus, to use a concept derivatively one attempts to match a 'foreign' example with an original example. The match can be more or less successful. Often the match is triggered by a desire that the foreign example replicate original examples. The dependence of derivative on original use can vary in character and complexity. But it always has to do with the particular understanding implicit in a given person's involvement in the practice.

Though derivative examples are dependent upon original examples, it is possible, none the less, for the former to affect one's understanding of the latter. Derivative examples might bring something new to the understanding of the concept under which they now fall. When derivative use affects the meaning of the concept in any of these ways, we react accordingly. Sometimes we decide or sometimes we come to accept (as participants in a practice) that we must expand or modify the meaning of the concept itself. And sometimes that persuades us to redefine the scope of the concept's original and derivative use. Neither of these reactions is problematic. On the contrary, it is precisely the possibility of expansion and modification of conceptual meaning (seen here in terms of the interplay of original and derivative use) that confirms the desired *openness* of many of our concepts.

---

[25] The way a musical work exhibits features is complex. It is appropriate to speak of a work as having such and such features if it is suitably packaged. If an adequate score is written, the music takes on a set of constitutive features determined by that score. Contrast this with the practice of improvisation, where musicians speak of a basic tune or rhythm, or refer to a performer's rendition of a tune, as the individuating principle. If a given improvisation is recorded and notated, this procedure effectively increases its number of constitutive features. Only too soon the improvisation acquires the status of a work and the work takes on the character of being composed. The improviser-jazz musician duly acquires the status of jazz composer.

This general view of conceptual use leaves open three more possibilities. First, a musician in 1810 and one in 1990, for example, might both function under the regulation of the work-concept, but because of possible modification of meaning, they might be working with a different understanding and with a different range of original and derivative examples. Whether or not this is the case, there is a dynamic and diachronic relationality, in addition to our affirmation or rejection of the past, linking together the successive stages of the concept, all of which preserves the concept's identity over time.

Second, to talk of original and derivative examples falling under a given concept does not mean this is the only way to classify the examples. It is possible to interpret and perform Beethoven's 'Spring' Sonata, say, in accordance with the ideals of a tribal rain dance. We would then be obliged—in the context of the corresponding practices— to describe the music not as an original example of a work, but as a derivative instance of a rain dance. If this example is too extreme, consider the reconception involved in our listening to classical works that have been jazzed up, used in films, or ranked in the Top Twenty.

Finally, one may speak of an object falling originally or derivatively under more than one concept. Musicians might deliberately produce music that fails to fall neatly under a single concept. Many musicians have recognized the limitations, say, of producing music solely under the dictate of romantic ideals and have chosen to employ notions associated with other types of music. Numerous composers have made use of folk-music and jazz, in order to make statements not fully comprehensible within, or explained by, a romantic aesthetic. To what extent, if at all, the work-concept has remained central to their musical production has differed from case to case.

IV

I have explained why, in historical terms, and how, in philosophical terms, the work-concept has come to have pervasive influence and extensive use. Now I am in a position to show what I promised to show much earlier on, namely, what effect all this discussion has on the philosophers' traditional use of the counter-example method.

Recall that analytic theorists had few qualms in using certain examples as examples of works. We saw how their theories were formulated on the basis of examples drawn from the repertoire of the early nineteenth century; their inquiries often began with a question like 'What sort of thing is Beethoven's Fifth Symphony?' That should no longer surprise us at all. But what, we asked before, are we to make of the fact that examples drawn from all sorts of music, but rarely from the nineteenth-century repertoire, have been appealed to, especially when a given theorist wanted to challenge someone else's theory or part of it?

The success of such a challenge—the success of the counter-example method—depended entirely, I suggested, on whether the chosen examples are works. The challenge rests on the assumption that the work-concept can be employed when speaking about types of music other than that of post-1800, work-music. If that assumption is well founded, then those who want to reject a purported definition of 'musical work' have before them an extremely broad range of possible counter-examples. Or do they?

If one adopts a traditional, essentialist approach to definition, or even a quasi-essentialist position, or even the view that the musical work is a closed concept, one will probably assume that all examples of works are such in virtue of their possessing certain essential or defining properties. Either an object of a given kind has these properties or it does not (in which case it is an object of another kind). As long as this basic tenet is held, the additional move to distinguish original and borderline examples in order to confine one's definition to the former is ultimately unnecessary. Strategically these additional moves might have a point, but all that matters in the end is our determination that a given object has or does not have the properties associated with the kind in question. If it does, it is entitled to serve as a counter-example to a relevant definition that excludes it. The counter-example method clearly operates on this view of things.

The counter-example method is called into serious question, however, when a more complicated account is embraced of what it means for something to fall under a given concept. Such an account was embraced when I argued that some concepts are open and therefore do not have their borders closed; when I drew a distinction between original and derivative examples; and when I suggested that something counts as an example of a kind not directly because it exhibits the appropriate properties but because there is an attempt to bring it under the concept by users of the concept within the context of a particular practice. Identifying counter-examples is now problematic, just because derivative examples cannot any more be used as such.

Because of the conceptual dependency that derivative examples have on original examples, objects come to count as derivative examples by virtue of the relation in which they stand to original examples. This sort of dependency implies that without original examples serving as the standard, there could be no conception of the non-standard. From this it follows that derivative instances can only be counted as works if the original instances of works have already been acknowledged as such. Now it makes little sense to think that a derivative example could be used as a counter-example to challenge either a definition confined to original examples or one accommodating, say, all currently existing works. To adapt a Hanslickian principle: no musical practice that is not guided originally by the work-concept can give the lie to a practice that is so guided.[26]

The way objects fall under open concepts is not just a matter of property possession. The *different* uses and the *dependencies* obtaining between these uses suggest that not all examples of musical works are classified as such in the same way and on the same terms. These complexities of conceptual use reveal the methodological basis of the counter-example method for what it is, and, more importantly, its limited use. But the full effect of this argument can best be seen if we bring to mind, first, the fact that not all challenges to concepts are of a narrowly philosophical nature. They are not all about the acceptance or refusal of purported philosophical definitions. Other sorts of challenge bring about modification and alteration in, or sometimes even the demise of, a regulative concept's force.

Imagine, then, an avant-garde creation that is exhibited as a work; it is received according to the *Werktreue* ideal. None the less, it turns out that the creator desired all along to challenge the work-concept in producing this creation. How are we to understand this challenge?

Suppose we offered a description of musical works based on those composed around 1800. If we subsequently wanted to accommodate examples of, say, aleatoric music, we would broaden our explanation. We would not, however, say the original explanation was incorrect or wrong in a strong metaphysical sense. Rather, we would say it was limited in scope. Composers who produce something designed to challenge the work-concept because they want to change, expand, or modify that concept do not suggest that Beethoven did not thereby compose musical works, though in a combative or polemical mood

---

[26] *On the Musically Beautiful*, 20: 'We see that vocal music, whose theory can never determine the essence of music, is moreover in practice not in a position to give the lie to principles derived from the concept of instrumental music.'

they might put the point in these terms. Rather, they think, Beethoven composed with an understanding of music they now believe to be based either on an ideologically unsound aesthetic, or on concepts too narrowly conceived or outdated. When musicians want to challenge a concept, they first have to acknowledge that the present meaning of the work-concept is as it is. Having done that, they then become involved in trying to expand or modify that meaning.

That a challenge to a concept depends upon an initial acknowledgement that the concept has a given, traditional meaning distinguishes this sort of challenge from that implied by the use of counter-example method. For it is one thing to broaden a concept's meaning as musicians want to accommodate more examples and new ideas. It is quite another thing to use new examples as counter-examples to philosophical definitions based on a certain range of musical works. Only the latter involves a claim that a given definition is metaphysically incorrect.

V

The preferred model of conceptual challenge presents difficulties not only for those who like to employ a certain philosophical method, but for all those who might like to think that such challenge is simply a matter of someone's proving or denying that a concept no longer has any meaning or force. For my model suggests two facets of conceptual change: first, those who wish to challenge a concept's regulative force usually find themselves paradoxically situated in a practice that is regulated by the very concept they want to challenge; second, that a regulative concept's alteration or demise is no less complicated a process than its emergence.

In the last decades, indisputable challenges and protests to the practice and theory of classical music have taken place. Such challenges have been presented in many forms and under various sorts of description. The most familiar are those presented as attacks on opus or concert hall aesthetics, on bourgeois and commodity aesthetics. But there are also attacks on opus-music that focus on its apparent élitist and gender-biased founding ideology.[27] Without intending to

---

[27] For useful discussions, see Adorno, *Philosophy of Modern Music*, tr. A. G. Mitchell and W. V. Blomster (New York, 1973); P. Bürger, *Theory of the Avant-garde*, tr. M. Shaw (Minneapolis, 1984), esp. 55–9; C. Ballantine, 'Towards an Aesthetics of Experimental Music', *Musical Quarterly*, 63 (1977), 224–46; and for more recent

do injustice to the subtle differences existing between these challenges, I find it generally unclear what the exact object or purpose of them has been, and whether or not they have been successful.

Cynical observers of the classical music world might say that whatever has been challenged, it has certainly not affected the force of the work-concept. Whatever the reasons, have they to do with survival, convenience, or compromise, no modern musicians (the ones the public know about) have brought about effective change in a work-based practice, for we still generally speak about music in terms of works. This cynical line can be pursued quite far until it flounders on the grounds of oversimplification.

Thus, it is not uncommon to hear of a tension existing between what musicians claim they want to do with music and what they actually do. This tension results in a basic paradox, which, as Tormey has written, 'has been exposed brilliantly, though . . . unwittingly by Cage himself'.

Cage remarked that the composer should strive to eliminate himself from his work and simply let the music happen. But where the composer succeeds in this project he must also fail, for in eliminating himself he ensures that it is no longer *his* music that he leaves us, and to the extent that it remains *his* music he has failed to extrude himself from it.[28]

Cage (not unwittingly I believe) speaks of having argued himself in theory out of a career which in practice he strives to maintain. The resultant paradox, which Renato Poggioli once referred to as 'the paradox of generations', is not Cage's fault.[29] What he and other avant-gardists demand in theory has to be contradicted by practice, if that practice functions with ideals one opposes in theory, yet, for whatever reason and however reluctantly, one accepts in practice. It is difficult to challenge in practice in a radical way that one at the same time participates in.

Conceptual challenge is complicated not only because it generates paradox, but also because it takes on different forms. Consider, first, the enormous expansion that has taken place in this century in our beliefs about what can count as musical. Such expansion has resulted

discussion, essays collected in Norris (ed.), *Music and the Politics of Culture* (London, 1989) and in Leppert and McClary (eds.), *Music and Society*.

[28] 'Indeterminacy and Identity in Art', 204.

[29] R. Poggioli, *The Theory of the Avant-garde*, tr. G. Fitzgerald (Cambridge, Mass., 1968), 35.

as much from the emancipation of dissonance and the rejection of tonality as from the introduction of new forms and instruments. All these challenges have altered the material of music, and therefore can be seen to constitute *material* challenges to the traditional concept of music. It is less clear, however, that these challenges have affected the packaging or individuation of music into works—say, the interrelations between performances and scores. Thus one can bring about material changes without thereby bringing about *formal* changes in a practice.

To assess whether formal changes in musical practice have occurred, one has to see whether there have been changes in the conception of notation, performance, creatability, autonomy, repeatability, artificiality, and product. And then one has to determine whether such formal changes, if they have occurred, have affected the force of the work-concept. Claims and rejoinders, then, that refer to the improvisatory role of the performer, or the lack of fixedness in compositions, are different in kind and effect than claims about material limitations imposed by the tradition.

Finally, there are challenges that are neither clearly formal nor material, but are framed instead as challenges to prevailing aesthetic, political, or social theories. These challenges are not mutually exclusive, since aesthetic theories, for example, have political as well as social consequence.

Take, now, Cage's declaration that his composition entitled *Music of Changes* is fully composed. That seems to be an endorsement of the work-concept, and so does his explanation. 'In other words,' he writes, 'though the sounds and then successions were to some extent dictated by chance, the notation is complete and must be observed by the performer.'[30] Were one to replace 'chance' with 'inspiration', little would distinguish this statement from one made by the most committed of romantic work-orientated composers. Yet when we hear Cage demanding that music be indeterminate and unintentional, when we hear him demanding that musicians let noise and aleatory elements into the concert hall, our assessment is apt to change. For what he seems to want to challenge now are formal notions of pre-composition, fixed instrumentation, repeatability, and compliant performance. Cage suggests the composer should 'give up the desire to control sound', and this looks like an overt challenge to the idea of a composer determining a work's sound-structure prior to its performance.

[30] *Silence*, 10.

But when we subsequently read that a composer should 'give up the desire to control sound, fixing clear his mind of music, and set about discovering means to let sounds be themselves rather than vehicles for man-made theories or expressions of human sentiments', it looks as if Cage's immediate aim is less to challenge the issue of compositional control, than to oppose an aesthetic claim that binds composers to their works via the relation of expression.[31] Now, it is no longer clear whether Cage intends to undermine the formal demand that the work be pre-composed, even if all that means is that cage specifies temporal limits for the performance.

Consider a claim once made by Karlheinz Stockhausen:

For me every attempt to bring a work to a close after a certain time becomes more and more forced and ridiculous. I am looking for ways of renouncing the composition of single works and—if possible—of working only forwards, and of working so 'openly' that everything can now be included in the task in hand, at once transforming and being transformed by it; and the questing of others for autonomous works just seems to me so much clamour and vapour.[32]

Clamour and vapour can pervade an atmosphere at one level and not at another. Stockhausen might well hold that autonomous, pre-composed works are antithetical to the sense of 'a musical experience being lived for the moment'. Still, openness and immediacy might none the less apply only to performances and not to works as such. The inclination to think this stems from the fact that composers can still maintain control over their works even if they do extremely little. There is a significant difference between performers who comply with a notation which specifies that they play what they will for as long as they want, and in so doing perform the composer's work, and performers who improvise freely around themes because that is what it means to produce their sort of music. What is absent in the second case, but what is present in the first, is the performer's commitment to *Werktreue*.

Recognizing that indeterminate performance or minimal specification does not necessarily imply indeterminate work, Cage and Stockhausen have sought ways to undermine their control over both performance production and their compositional procedure. If the 'work' itself can be formed as a result of as little controlled interference as possible, the resulting music is as close as it can be to being uncomposed or

[31] Ibid.
[32] K. H. Wörner, *Stockhausen: Life and Work*, tr. B. Hopkins (Berkeley, Calif., 1973), 110–11.

unintentional. Now it looks as if the notion of the fixed artifice implicated by the work-concept is under threat. But still the threat is complicated.

Cage's most famous 'work', *4' 33"* (1952), has long been recognized as ironic in its description as a 'work of silence'. The music heard within performances is constituted not by silence but by the random sounds heard by any individual person who happens to hear those sounds within the designated time. Cage's 'work' reflects an attempt to shed music of institutionalized constraints imposed by composer, performer, and concert hall. He aims to bring music back into the real or natural world of everyday sounds. All sounds, in his view, are musical.

Despite the apparent absence of pre-determination in the composition, despite the experimental nature of performances where the emphasis has deliberately been placed on natural sounds rather than on intentionally produced sounds, Cage has not obviously succeeded with *4' 33"*, and other such 'works', in undermining the force of the work-concept *within* the musical institution. First, he has maintained control (however minimal) over the music. It is because of his specifications that people gather together, usually in a concert hall, to listen to the sounds of the hall for the allotted time period. In ironic gesture, it is Cage who specifies that a pianist should sit at a piano to go through the motions of performance. The performer is applauded and the composer granted recognition for the 'work'. Whatever changes have come about in our material understanding of musical sound, the formal constraints of the work-concept have ironically been maintained.

Did Cage come to the compositional decisions that he did, out of recognition that people will only listen to sounds around them if they are forced to do so under traditional, formal constraints? Does he do what he does to maintain his career as a composer of musical works? Remember, no composers, performers, and audiences are needed once individuals learn how to listen (in Cage's ideal terms) to natural sounds as music. Apparently, Cage endorses the formal packaging of music within the context of a concert hall for precisely these reasons. He regards himself both to be a teacher and to be working within the tradition of classical music.[33]

---

[33] Apparently Cage maintains his career as a composer of musical works partially to honour a promise once made to his teacher, Schoenberg. I have benefited from R. Kostenlanetz's discussions of Cage in his *On Innovative Musicians* (New York, 1989), 21 ff.

Contrary to the spirit of an anti-survivalist aesthetic, in which music is performed for the present and is not composed to last, Cage's 'works' have survived and have come to be representative, no less, of the avant-garde repertoire of musical works. Performers and listeners receive his music with the same respect many of Beethoven's contemporaries had for his innovative works. In both cases, such respect involves feelings both of awe and bewilderment. Cage's 'real' and 'random' sounds have not stayed real or random. The 'real' sounds of 'his' 'work' have been made subject to all the traditional, temporal, presentational, organizational constraints associated with any concert hall performance. Cage has not always, as other artists have more often done, challenged the work-concept by taking music outside the institution onto the streets. He has tried to bring the 'outside' world back into the institution, but apparently to no avail. The institution has not shown itself to be so flexible that it could allow Cage's 'real' sounds to remain real.[34]

This conclusion does not point necessarily to Cage's individual failure. It points to the paradox many composers face when they attempt to produce something revolutionary from within the institution. Despite all their claims that their music should not be seen in such and such a way, they continue to see it treated in just that way. The paradox seems to be inevitable. For as soon as musicians claim to be, and then become accepted as, participants in a specific practice of music-making, the temptation of the mainstream is not to see the challenge they present to the practice as a radical challenge to their practice. More often, the mainstream will interpret and then accommodate the music to suit themselves; more often, they will simply pretend that the music does not exist.

VI

Appearances notwithstanding, the paradox described does not mean that challenges to the musical tradition have had no effect on the force of the work-concept. As we begin to explain why this is so, the premises of the cynical argument offered in the previous section begin to unravel, though in the end the conclusion remains. Thus an

---

[34] 'The *objet trouvé* . . . unlike the result of an individual production process but a chance find . . . is recognized today as a "work of art" ', Bürger argues in a similar vein. 'The *objet trouvé* thus loses its character as anti-art and becomes, in the museum, an autonomous work among others' (*Theory of the Avant-garde*, 57).

adequate account of conceptual challenge cannot assume (as I did just now for the sake of the argument) that one can assess formal challenges independently of material ones, or even independently of challenges to aesthetic theory. Conceptual challenge is more complicated than we have supposed so far.

Challenge to a regulative concept takes many forms, as we already know. It also functions ambiguously, for normally one cannot tell whether the intent of a challenge is to modify or alter the particular force of a regulative concept or to undermine its force altogether. The challenge is further complicated by the possibility of constant change in a concept's meaning. Regarding the latter, a regulative concept acquires its force through its association with a whole range of interrelated concepts and theoretical claims. Yet not every challenge to one of the associated moments necessarily challenges the force of the concept. This is because the relation between the regulative concept and the practice and theory with which it is associated is not static. Even with broad knowledge of the work-concept's emergence, it remains an open question whether the force of that concept has continued, since its birth, to depend on the explicit acceptance of the material, formal, or aesthetic assumptions with which it was originally associated.

It is possible, in other words, for concepts to shed their histories by taking on contemporary meanings. It is possible also for concepts to become neutralized regarding their commitment to any specific set of assumptions. Neutralization does not imply absence of content; it implies a lack of strong allegiance to too specific an ideological, political, or (where applicable) an aesthetic content. Neutralization might imply the presence of a new content extrapolated from a range of changing assumptions. It might point to the fact that what was once a concept with very specific content has come almost to look as if it were generic. If it turned out, now, that the musical work-concept had become neutralized, that might mean that its force had effectively freed itself from its original romantic associations. And if it did mean that, one could now understand how musicians could reject outright the romantic aesthetic without thereby rejecting the contemporary work-concept.

Very briefly, one movement and one development in the history of music could be interpreted as having neutralized the work-concept of its romantic associations: formalism and mechanical reproduction. Central to formalism is an overriding emphasis on the well-formed, self-sufficient work, whose material and form are united such that even the relation of expression joining composer to product is overridden by the demand that one looks only to the work itself. 'The work speaks for itself, even without the name of the composer,' Schumann once wrote. A century later Nadia Boulanger claimed that, for her, 'the greatest objective is when the composer disappears, the performer disappears, and there remains only the work'.[35]

Formalism has increasingly inclined towards positivism to the extent that reference to 'subjective' principles having to do with personal, 'pathological' feelings has effectively been banned from the account. Such anti-subjectivism has found further support in technological changes affecting forms of musical production and reproduction. The invention of a technical means of production undermines, and imposes limitations upon, the contribution of human participation in the sphere of musical production, or so it has been argued. Through electronic equipment, both notation and performance can be given a form approximating more closely than ever before to the condition of the work itself, because such equipment greatly reduces the possibility of human error.

Both formalism and technology have had an undeniable impact on musical production. Yet neither, to my mind, has severely or obviously undermined the romantic basis of the work-concept. Despite the emphasis on a work's self-sufficient unity, little in formalism recommends denial of the romantic aesthetic. Recall that in the nineteenth century, formalism was a strong current within and not opposed to romantic thought. It is not surprising, therefore, that Hanslick, for example, supported many if not all the basic tenets of the romantic, musical aesthetic. More surprising perhaps is the fact that many modern formalists of this century, such as Milton Babbitt, have also endorsed, more or less opaquely, romantic ideas about the status of artists and their works.[36] And the same conclusion can be drawn regarding technological developments. They have as much reinforced and fetishized the romantic (some say essentially bourgeois) conception of composer, performer, and conductor, as they have dehumanized musical practice in a negative sense. Because formalist theory and

[35] *Schumann on Music*, 125. Nadia Boulanger, in A. Kendall, *The Tender Tyrant: Nadia Boulanger* (Wilton, Conn., 1976), 115. Boulanger finishes the sentence with the proviso: 'but that you can do only with great masterpieces. They stand by themselves, only by themselves.'

[36] 'Who Cares if you Listen?', *High Fidelity* (1958), rpt. as 'The Composer as Specialist', in R. Kostelanetz (ed.), *Esthetics Contemporary* (Buffalo, NY., 1978), 280–7.

technological development have done as much to reinforce, as they have to modify, the connection between the work-concept and major tenets of the romantic aesthetic, the conclusion that they have straightforwardly neutralized the work-concept of its romantic associations cannot easily be sustained.[37]

Let us return to recent avant-garde practice, for maybe musicians are using the work-concept here without its original, romantic content. Earlier, when discussing Cage's position, perhaps we did not sufficiently recognize the extent to which he has come to use the term 'work' loosely. Perhaps he uses the term merely to suggest an occasion for a certain kind of musical performance. Cage often refers to 'happenings' or 'chance' music, and there is every reason to believe that he would be perfectly happy to cease to use the word 'work' altogether. Cage is not the only one to have spoken in this way. Other musicians have spoken of musical occasions, sound environments, and musical situations just to differentiate their music from the tradition.[38]

When musicians bring about happenings outside concert halls, when they suggest audiences bring instruments to play at musical occasions so as to be performers as well, when they refuse to produce scores or recordings to foreclose the possibility of the music's surviving past the occasion, when they stress that a performance is an end in itself and not merely a channel through which a work flows, and when they impose no temporal limits on a performance, we do, after all, seem to be observing direct challenges to the traditional work-concept. For not only do the musicians say they are challenging the concept, but their directions are also explicitly intended to undermine the traditional, formal conditions associated with work-based music.

---

[37] Modern formalism of this century has often been expressed in terms of the separability principle, as a doctrine that sustains the view that art exists for its own sake, and should not be subjected to the mundane forces of a commercial musical market. See R. Williams's discussion of Russian formalism in his *The Politics of Modernism: Against the New Conformists*, ed. T. Pinkney (London/New York, 1989), 164 ff.

[38] I am thinking, for example, of Allan Kaprow's 'happenings' of the 1950s. For discussion of happenings, see Dick Higgins 'Intermedia', in Kostelanetz (ed.), *Esthetics Contemporary*, 186–90. Higgins writes: 'the Happening developed as an intermedium, an uncharted land that lies between collage, music, and the theatre. It is not governed by rules; each work determines its own medium and form according to its needs. The concept itself is better understood by what it is not, rather than what it is' (189). In part, he argues, it was produced in reaction to the mechanistic division between performers, production people, a separate audience, and an explicit script. See also W. Mertens's excellent discussion of the challenges of minimal music and happenings to the work-concept in *American Minimal Music* (London/White Plains, New York, 1983).

Recall, for example, Neuhaus's reason for moving his acoustic sound environments far away from the museum arena. Having installed in 1983 an acoustic system in a sculpture court at the Whitney Museum, Neuhaus judged the action unsuccessful. The problem was that 'most people went out and *tried* to hear it', he said, because it was 'in the context of a museum exhibition'. As Calvin Tomkins further reported the story, Neuhaus wanted 'people to come upon his sound accidentally and respond to them without preconceived ideas or expectations'. Neuhaus subsequently took his installation elsewhere.[39] And Cage and many others have done and said all the same things in the same spirit and with similar intent. So why does it remain so difficult, as I believe it does, to assess the import and impact of their challenge?[40]

It remains difficult for all the sorts of reasons that we gave earlier, plus one more. Musicians do not always speak of their music in terms of occurrences, but continue to use all the terminology, as well as engage themselves in the practices, associated with work-production. Because of this, it is not always possible to determine whether they are doing what they do in a way wholly or partially neutralized of a commitment to an aesthetic and form of production of which they no longer approve, or whether they are using the work-concept and all that it entails with a wholly new understanding. How helpful was it when Marcel Duchamp, for example, inquired in 1913 into the possibility of producing works of art that are not works of 'art'? Most persons understood what he meant by 'work of "art" ', but what did he mean by 'work of art'?[41]

---

[39] C. Tomkins, 'Onward and Upward with the Arts', *The New Yorker* (24 Oct. 1988), 110–20. Recall also Satie's 'furniture music', for which he gave the explicit instruction that one should not listen, but keep talking. Recently reported by Roger Shattuck, 'Mascot of Modern Music', *New York Review of Books* (15 Mar. 1990), 32.

[40] In 'On the Decline of the Concept of the Musical Work' Dahlhaus argues that the work-concept found its force more as an aesthetic ideal than as an ideal affecting the everyday activities and packaging of music and musicians. For him, it follows that the concept's decline rested on the failure for practice to fulful its 'aesthetic claims'. Thus, he writes, the recent polemics against the concept of the musical work 'are directed against a dominant and yet powerless idea' (*Schoenberg and the New Music*, 221). My argument suggests, by contrast, that the concept remains both dominant and powerful.

[41] *The Writings of Marcel Duchamp*, ed. M. Sanouillet and E. Peterson (Oxford, 1973), 74. Recall Adorno's equally 'paradoxical' remark that '[t]oday the only works which really count are those which are no longer works at all'. *Philosophy of Modern Music*, 30.

No one doubts that it is generally more convenient and pragmatic to continue using familiar terminology. It is more convenient because it provides a connection with the practice which one wants both to be part of and, at the same time, to change. Thus the paradox arises again. There is nothing negative about one's working with this paradox; on the contrary, it helps provide for dynamism and change in a practice. The only problem is that the convenience of using traditional terminology usually works against the revolutionary force, and gives a place to revolutionaries that is more firmly within than clearly beyond the status quo.

This last fact points to yet another aspect of the paradox inherent in the avant-garde movement. While the movement puts the status quo into question by virtue merely of its presence as an antagonistic 'minority' or 'marginal' culture, it none the less constitutes, at the same time, part of what defines the culture as a whole. To bring about change in the status quo entails, then, change not only in the mainstream but in marginal cultures as well.[42]

This whole argument also applies to cases where the work-concept is employed in practices not producing European, classical music. The story I told earlier about conceptual imperialism might well be thought to underestimate the extent to which non-classical musicians have borrowed terminology without taking on an entire package of beliefs and ideals. Still, I believe that if forms of imperialistic domination motivated many musicians to adopt work-terminology in the first place, it was when they and their music subsequently acquired either 'respectability' or control of the market that were they able to use the terminology as if it were neutralized or devoid of its romantic content, when and if that was desired—thus the popular move to 'roll over Beethoven'.

Like the case of the avant-gardist's continued use of the terminology, one has, however, to be able to distinguish between what has been intended and what has come to be understood in musical society at large. For when one ignores the fact that there might be a great discrepancy between the two, misunderstanding as to the musician's commitment is likely to prevail. Wynton Marsalis's appearance as a jazz musician playing in the concert hall, clothed in classical garb, might just be an appearance. It might just be strategically misleading, or, then again, it might not be. Sometimes, it is hard to tell.[43]

---

[42] For comparable discussion, see Poggioli, *The Theory of the Avant-garde*, 108 ff.

[43] A recent article by Peter Watrous in the *New York Times* (10 Mar. 1990, Arts

VII

Dismantling the force of a regulative concept requires more than overthrowing or negating the original moments of its birth. Where a concept has separated itself from those moments through neutralization or any other sort of change, we have to challenge it also on contemporary terms in order to dismantle it. Small-scale challenges are inadequate to the whole job. To fully dismantle the force of a concept, one needs a global paradigmatic shift, of at least equal complexity to the shift that founded the force of the concept in the first place.

What might a global change to our work-based musical practice look like? There are answers that reflect those which recent critics of modernity have given, all of which are found also in recent musical writings. Thus, for example, perhaps we should return to a condition of musical practice that we had once before. Perhaps we should use a mythified and idealized past to legitimate a new vision of the future. Perhaps classical music should try to approximate to the condition of non-western music or western music of a non-classical sort to help us find or increase our resistance to the work-concept.

Returning to or rediscovering conditions of the past, even attempting through philosophy, religion, or aestheticism to transcend the limitations of our times, even forming exclusive clubs within counter-cultures to make new music available only to the few who already understand and sympathize with it, all serve as temporary and literally marginal solutions, but they do not effect sufficient change. Sufficient change is global and depends upon our moving forward with the understanding that paradigm shifts can and do occur within one's own culture or practice. Global change has to be internally generated as much in the mainstream as in the margins of culture. It results from new desires, new beliefs, new theories, and innovative and provocative experiments of imagination, all of which have to bring about a general feeling of discontent with the old.

Section, 14), begins with the thought that '[alt]hough it's not often brought up, many jazz fans harbor the nightmare that jazz will become like classical music, cleaned up and locked away in the museumlike confines of American concert halls. Jazz, whose backbone is improvisation, would no longer be improvisatory. Its creative urges would dry up and the form would blow away. . . . And there is cause for such concerns.' Watrous subsequently explains that Wynton Marsalis, with the latter's deep respect for Blues history, is 'intent on making sure the musical high points of American culture remain vital and accessible'.

Only with global changes are paradigms or movements brought to an end. Only when movements are brought to an end, can they be fully assessed. Hence, we will only be able to fully assess current, avant-garde challenges to the work-concept when the challenge is, as it were, no longer required. But that will not be for a long time as far as I can tell. For at present, the establishment remains largely content with the familiar shape of musical production. Publicly expressed discontent is focused on issues of government support of the arts and the lack of adequate funding for music education. Few express discontent with the basic way of looking at classical music, but that is because people are still generally unaware of how they do look at this music. The mainstream thus remains committed to the work-concept, and furthermore, it seems, to its traditional, romantic force and meaning.

To be sure, not everyone is content, even in the mainstream. Some musicians have thought that *Werktreue* and all that it implies in the practice of music has retained a meaning no different from the most romantic, autrocratic, and most imperialistic of ideals. Alfred Brendel argues that the ideal is antiquated, fetishistic, fallacious, and bourgeois. '*Werktreue* smacks of credulous, parade-ground solemnity,' he writes. It has, he continues, the connotations of 'Viennese Classical Training' and 'Nazi slave mentality'.[44] But even such extreme sentiments have not generated significant change.

Even where tension over the work-concept has arisen, solutions have been found to satisfy musicians such that they could continue to act and think about music in much the same way as before. This becomes evident as we turn finally to look at two particular 'mainstream' debates over the ideal of *Werktreue*. These debates reveal that the work-concept contains enough subtle points of indeterminacy, stemming from indeterminacies in underlying theory, for musicians to be able to use the concept more to their advantage than to their disadvantage. And that this is the case might ultimately account for the long-term success of the work-concept in the field of music.

Of course, the long-term success of the work-concept is not necessarily a bad thing. To establish that it was, one would have to show that the work-concept and the practice it regulates are wholly bad all

---

[44] Brendel, 'Werktreue—An Afterthought' in *Musical Thoughts and Afterthoughts* (Princeton, 1976), 26–37. Adorno discusses the problem indirectly in his essays comprising *Introduction to the Sociology of Music*; see also his 'Bach Defended against his Devotees', *Prisms*, tr. S. Weber and S. Weber (Cambridge, Mass., 1967), 133–46.

the way through, and that cannot be the case in any straightforward sense. However, there is one claim made by supporters of the work-based practice that we most surely need to revise—the claim as to the work-based practice's universal and absolute validity.

VIII

In the previous chapter, I described a way in which performers reconciled themselves to subservience to the composer and the *Werktreue* ideal by introducing a parallel practice of virtuoso performance often based on extemporization. Following that description, I mentioned that conductors had not been able to find such a parallel. But, I said, their position was none the less not an unhappy one. I wish now to explain that.

The cynical explanation is that conductors—at least the successful ones—earn much greater salaries than all other classical musicians—except virtuoso performers. The serious explanation concerns their continued allegiance to the ideal of fidelity. Modern conductors have found themselves in an uneasy position because they are simultaneously regarded as masters and servants. They are regarded as masters and leaders of orchestras and of the art of interpretation. As such, they are accorded the reverence normally reserved for saints—or composers. But they are also regarded as servants to the composer's works. Now they are apparently comparable to 'coach drivers' and 'midwives'. In musical terms, conductors are asked to show fidelity to the works, yet offer, at the same time, independent, novel, and personal interpretations of them. They are asked to balance fidelity to the work specified by the composer against their allotted freedom for artistic gesture. How much variation can there be in interpretative gesture before fidelity is undermined? With ambivalence in mind, conductors have come to the conclusion, as William Steinberg once put it, that there is 'no function in the entire realm of the performing arts as universally misunderstood as that of the conductor'.[45]

A resolution might be found were one to emphasize one role, master or servant, at the expense of the other. One might choose to defend this move on sophisticated grounds provided by recent hermeneutical theories of interpretation. Within this more general framework one would soon see, however, that the ambivalence of the conductor's art

---

[45] Galkin, *History of Orchestral Conducting*, p. xl.

is shared by any agent involved in interpretation of the arts. And even were we to confine the inquiry to the musical domain, we would find musicians other than conductors coping with the same ambivalence. Pierre Boulez once expressed (perhaps inadvertently) the dilemma of all musical agents, except composers, through the insightful metaphor of a mirror.

The listener is only a reflection of the orchestra and a work in performance is built like a telescope. The focus is the work, and a conjunction of mirrors is used to magnify—first the conductor is mirrored by the orchestra, then the orchestra, is mirrored by the audience.[46]

With these words, Boulez pointed to the problem of subservience arising from the romantic stipulation that all activities be transparent in order to let the work shine through.

Since 1800, power relations generated by the work-concept have been most apparent. Performers, for example, who have not made it as independent virtuosi have had to temper their individuality and freedom to serve not only the music they perform but also the conductor they perform under. Orchestras have even been spoken about mechanistically, not as groups of individual musicians, but as instruments in the hands of conductors. 'Wagner controlled the orchestra as if it were a single instrument and he was playing it.' Pierre Monteux spoke of orchestra-players as 'cogs in the wheel'. These metaphorical descriptions clearly undermine the equality of conductor and orchestra reflected in the mirror metaphor. Yet, as power relations naturally operate, orchestras manage on occasion to assert themselves. 'Don't forget that your authority is delegated,' they remind their conductors.[47]

Whether conductors use orchestras as mechanical vehicles or whether they help to bring out the expressive potentialities of players is an issue with a distinctly moralistic flavour. One could ask whether it was possible, given a Kantian or humanitarian prescription, for conductors to use orchestras both as a means and as an end, but never just as a means. Ideally, the answer would be positive. Pragmatically, it would probably depend on whether conductors had resolved their own feelings of ambivalence.

[46] In R. Jacobson, *Reverberations: Interviews with the World's Leading Musicians* (New York, 1974), 29.
[47] Wagner quotation from Horowitz, *Understanding Toscanini*, 41; Monteux's warnings printed on the inside cover of Galkin's *History of Orchestral Conducting*.

As complex power relations cause tension between conductors and orchestras, so they cause even more tension between conductors and composers. Early on, Gounod reminded the conductor that he was

> nothing more than the driver of the coach engaged by the composer. He should stop at every request or quicken the pace according to the fare's orders. Otherwise the composer is entitled to get out and complete the journey on foot.[48]

The power relations between composers, conductors, performers, and audiences actually mirror nothing other than the more elusive, theoretical struggle implicit in the relations that hold between works, performances, and audiences. That is to say, whenever musicans discuss the merits of treating each other in one way or another, eventually they are forced to ask the modern question: wherein resides the meaning and value of a musical work—in the work itself, in its realization through performance, or in the interpretative act of listening to a work?

If a work has meaning independently of its performance or interpretation, composers merely require willing and able performers and conductors to reveal the meaning of their works. Contrarily, if works acquire meaning through interpretation, then conductors and performers have a more important and independent role, perhaps even a role that undermines the demand for fidelity to the composer's work. For if works have no meaning in themselves, what remains for conductors to be faithful to, other than a set of empty signs to which they then give meaning? And finally, if the meaning of a work is formed through interpretative acts of listeners, the actions and roles of composers, conductors, and performers have to be defined in such a way as to give the listeners completely free, interpetative rein.

Despite the theoretical alternatives, mainstream conductors have not been convinced that they should dispense with the ideal of fidelity to composers and their works. Consider two notions which have been amply used in the literature on conducting: interpretation and correct rendition. Both are embraced under the more general notion of a faithful performance and both have been described to indicate the

[48] J. Harding, *Gounod* (New York, 1973), 122–3; taken from Gounod's complaint to Count Walewski about his (Gounod's) lack of control over the performance of *La Reine de Saba*. 'On foot!', replied Walewski, '[p]eople already complain that composers don't do the journey fast enough. Stay in your coach, M. Gounod, and I will try to make the driver see reason.'

objectivist's belief that a work is fixed in meaning before interpretation takes place. Objectivity, after all, is the driving force behind *Werktreue*.[49] Both notions are thus described in terms of fidelity to a composer's intentions and the work's score. Hence, though the connotations of 'interpretation' and 'correctness' might well be thought to differ, the use of these words in the musical world has been effectively interchangeable.

Is the conductor's claimed ambivalence only apparent then? After all, if interpretation and correctness are both understood in terms of fidelity, conductors remain by their own admission servants to composers and their works. Before drawing any conclusions, let us see how conductors view their own situation. The most extreme statement comes from those who say that though the *Werktreue* ideal has exerted a power over conductors from the beginning of their profession, it no longer should. Then, there are those conductors for whom the sort of praise that Toscanini received as 'the priest of beauty, the consecrated celebrant, abstracted, absorbed, awaiting gravely the trembling of the Temple's veil' seems entirely appropriate.[50] With a description like this in mind, one begins to understand why, as Daniel Barenboim once suggested, audiences sometimes believe that the conductors are actually playing the music. Some might even want to believe that conductors compose the music as well.[51]

But conductors who like to hear such descriptions of themselves often proclaim complete submission to composers as well. Leonard Bernstein wrote:

Perhaps the chief requirement of all is that [the conductor] be humble before the composer; that he never interpose himself between the music and the audience; that all his efforts, however strenuous or glamorous, be made in the service of the composer's meaning—the music itself, which, after all, is the whole reason for the conductor's existence.[52]

Liszt also proposed that 'the genuine task of a conductor consists in making himself, manifestly superfluous'.[53]

---

[49] Cf. Horowitz, *Understanding Toscanini*, 98 ff.   [50] Ibid. 101.

[51] Barenboim in Jacobson, *Reverberations*, 16: 'Today, conducting is a question of ego: a lot of people believe conductors are actually playing the music.' Cf. Adorno's remark that 'what serves unreal ends appears as if it were real, and the conductor acts as if he were creating the work here and now' (*Introduction to the Sociology of Music*, 105).

[52] *The Joy of Music*, 156.

[53] Liszt, 'A Letter on Conducting' (1853), *Gesammelte Schriften*, v, ed. L. Ramann (Leipzig, 1880–3), 232: 'die wirkliche Aufgabe eines Kapellmeisters darin besteht, sich *augenscheinlich überflüssig* zu machen.'

Most conductors end up adopting the position that while committed absolutely to the ideal of fidelity, they none the less recognize the importance of interpetation. Berlioz summed up the position when he wrote that the primary responsibilities of conductors are 'fidelity to the directions of the composer' *and* 'communication of personal conviction, intensity and spontaneity'. But the overall idea, he concluded, 'is to direct the study of a work unknown to the performers, clearly setting forth the author's conception and rendering it salient and distinct, obtaining from the band that fidelity, unity and expression without which there can be no real music'.[54]

What, now, of the solution that decrees that there is no genuine ambivalence in the conductor's role because conductors remain willing servants to composers? Unfortunately, this solution still cannot be too quickly pulled out of the hat. For, given the merest hint of an understanding of interpretation framed by something other than the demand for work-fidelity, the conductor's problem re-emerges in full force. As soon as conductors try to free themselves from the burden of fidelity, the question arises as to how they will celebrate their newly found independence in musical performance. Just how will they constrain their own creativity? To what extent will they still be bound by composers' scores? What will guide their interpretations of a given work?

In the end, the ambivalence conductors feel does not arise through any feature peculiar to their profession, but because neither they nor anyone else has determined the nature and value of fidelity and the space left for interpretation within the confines of fidelity. And no one has done this because no one has answered satisfactorily the basic question as to what value fidelity or interpretation should have and to whom either should be of value. Should fidelity tell us something about the value of musical work itself, the performance, or the listener's aesthetic experience? Should fidelity be of value to the conductor, the composer, or to the audience? Even when musicians have spoken of all activities as being directed towards the work itself, few have been happy to conclude that all value is exhausted by a given work, even if it is the source of all musical value. Most have wanted the performance and reception of works to have intrinsic value as well.

Unfortunately, the limits and meaning of fidelity will never successfully be derived from some purported proof that a given way of

---

[54] Galkin, *History of Orchestral Conducting*, 583 and 276.

thinking about something is absolutely the right way. There is nothing about the concept of a work, the relations between works and performances, or works and scores, or works and experiences of them, that is going to tell us where the locus of musical meaning 'really' resides. There is no muse to appeal to. All we have are complex theories, and the practices to which these theories become attached; and these theories never become so well worked out that they provide all the answers.

And there is never just one sort of theory to consider. Apart from romantic theory, there are numerous theories of interpretation, many of which are currently challenging demands for fidelity and authenticity in the arts. That these theories are starting to be taken seriously might prompt the feeling that musical understanding must gradually change. If change is not desired, at least we might learn to accept another fact, that competing theories, agents, and even ideologies are indispensable to a healthy, living and changing practice, and that such competition should be encouraged and not seen as problematic. The tension existing between fidelity and unconstrained interpretation is a healthy one. For what actually results from such competition is a wealth of marvellous and very distinctive performances of musical works. Perhaps the tension between musical agents is healthy for that reason as well.

If we have a tradition of interpretative activities and ideas which, despite constant claims to fidelity or correctness, continuously undergo subtle revisions, should we still mind that conductors could claim, were they so inclined (and as a critic present at a Toscanini perform-ance once did), that each or any of their performances had actually become the work itself? Claims that point to someone's having got something absolutely right, and claims that reveal a belief in the absolute rightness of an ideal, do matter. Potentially, though maybe less so in the musical world than in the world at large, adopting an absolutist or rigidly objectivist position based on the hegemony of certain beliefs and ideals is dangerous. Actually, Elias Canetti nearly went so far as to conclude that the conductor displayed some char-acteristics of the political dictator.[55]

What absolutists do not generally see is that nothing important is lost by dispensing with absolutist claims, except the general feeling of

---

[55] E. Canetti, *Crowds and Power*, tr. C. Stewart (New York, 1962), 394–6. Cf. Adorno, 'Conductor and Orchestra', in his *Introduction to the Sociology of Music*, 104–17; and Keller, *Criticism*, 21 ff.

security that certain particular actions are 'really' worth undertaking. Without absolute grounds, structure and ideals remain, as does the possibility of finding well-founded criteria of truth, meaning, and value. Ideals function even in the absence of an absolute foundation. They function not least because it is thought valuable for them to do so, and that value can be explained and supported in numerous ways. In the end, musicians must just ask themselves whether the most satisfactory form of musical criticism is one that is based on the ideal of *Werktreue*. If it is not, they must seek an alternative. No musician is necessarily bound to this ideal, however pervasive and persuasive the romantic aesthetic.

The present situation, however, is such that, in the absence of complete theoretical determinacy, conductors survive with their double role. Making use of essential gaps of theoretical indeterminacy in the work-concept has provided them with a way to reconcile conflicting desires and descriptions of their musical duties. Whether or not this is a good thing, conductors really have no reason to complain. The only complaint that is justified is when any conductor, or any other musician in fact, fails to recognize the ramifications of their being regulated by the work-concept, and when, in *naïveté*, musicians demand that performing and listening to music always means doing the same kinds of things. For this demand too often stems from a belief that *Werktreue* has somehow proved itself to be absolutely the right ideal to have.

IX

These remarks on the profession of conducting are relevant not only to activities whose rationale developed according to the ideal of *Werktreue*, but also to activities that are necessarily antagonistic to the pervasive force of that ideal. This is evident as we turn, finally, to consider the current debate surrounding what has come to be known as the authentic or historical performance, or the early music movement.

The movement originated in the second half of the nineteenth century in reaction to the imposition of perpetually modernizing performance techniques upon music written before 1800. Its historically informed aim was to 'authenticate' the performance of early types of music by performing the relevant compositions on 'original' or 'period' instruments, as they were performed at their time of composition. That involved, according to its proponents, replicating in performance the music's original historical context of production in the spirit of playing the music as it 'really' or 'authentically' was.

In recent years, the accreditation of authenticity to historically informed performances has generated heated debate. Those who have objected to the movement have done so not because they dislike the performances, but because they have had theoretical reservations which, to their minds, cannot be ignored, given the more than usually extreme incompatibility between the expressed theory and actual practice of 'authentic' performance.[56] What is criticized most often is the severe indeterminacy in the proclaimed ideal of authenticity. As we shall observe in the following brief remarks, the ideal seems to have been understood in three ways, as fidelity to original conditions of performance (especially to the instrumental conditions), as fidelity to music (perhaps to the *Klangideal*), and as fidelity to the work.

A basic difference is first recognized to exist between early and modern conceptions of the role and value of instruments and performing techniques. This difference has prompted questions regarding the extent to which early musicians concevied of their music with know-ledge in mind as to how it would be performed on the occasion for which it was written, and what that meant if they did. Did early musicians account for the fact that instruments had specific physical capabilities? Did they accommodate the fact that performing conven-tions had elements which perhaps interfered with what they might have thought to be an ideal performance? Consider—to borrow Charles Rosen's examples—the stand-tapping or intermittent continuo-playing of the eighteenth-century conductor. Musicians recognized that such activities were necessary to secure a successful performance. But did that recognition entail that they regarded them merely as

[56] One should not confuse the theoretical debate between serious proponents and critics of the movement with theoretically clothed reservations expressed by the media—by those who continue to support 'modern' performances, or by those who perhaps dislike a recording industry that has exaggerated the claims of the movement to sell more records. Words like 'authentic', 'original' have unparalleled persuasive and marketable force. I have based my remarks on the following body of literature: Adorno, 'Bach Defended against his Devotees'; L. Dreyfus, 'Early Music defended against its Devotees: A Theory of Historical Performance in the Twentieth Century', *Musical Quarterly*, 69 (1983), 297–322; H. Haskell, *The Early Music Revival*; N. Kenyon (ed.), *Authenticity and Early Music: A Symposium* (Oxford, 1988); J. Kerman, *Contemplating Music: Challenges to Musicology* (Cambridge, Mass., 1985), ch. 6; M. Morrow, 'Musical Performance', *Early Music*, 6 (1978), 233–46; C. Rosen, 'Should Music be Played "Wrong"?', *High Fidelity*, 21(1971), 54–8; and R. Taruskin, *et al.*, 'The Limits of Authenticity: A Discussion,' *Early Music*, 12 (1984), 3–25, 523–5.

extraneous 'necessary evils' of performance, or as forming an essential part of the music itself—the actual *Klangideal*?[57]

As these questions are commonly posed, it is fairly assumed that the answers given to them should be accommodated in subsequent authentic performances. Were it established, for example, that a musician composed not just for a specific kind of piano but intentionally for one actual piano, and were one to believe that authenticity meant one's being faithful to the composer's intentions, an authentic performance would be one in which that very instrument and no other was played. However straightforward the reasoning, critics have rarely seen it followed through in practice. The resulting tension might well have proved a convenient practical loophole for authenticists, but it has been a major source of exasperation for the critics.

Thus, critics have argued against authenticists that their interpretations and uses of the past are selective, that they choose to conform to some but not all conventions of the original performance practice—only to certain aspects of the original sound, instruments, acoustic setting, only to certain dynamics, embellishments, and styles of vibrato. Critics have wanted authenticists to admit their failure to meet the ideal, or if not that, to argue themselves out of the commitment that to be completely and not just selectively faithful to early music one has to meet all the original conditions. Compositions, where appropriate, must arguably be played on the 'right' days, marking the original occasions. Original, acoustic settings must be replicated, as must pitch levels and even performers' costumes.[58] For perhaps such things once were, and therefore remain, essential to the music performed.

Even where historical authentication has been recognized to be necessarily selective and partial, even where authenticity has been recognized to be an ideal of performances and therefore not perfectly attainable, the charge remains that the ideal of historical authentication is problematic in itself, rendering irrelevant all subsidiary problems. Thus, critics have argued that the ideal mistakenly assumes that we can rid ourselves of our modern ideas and by a leap of imagination return to

---

[57] Rosen, 'Should Music be Played "Wrong"?', 55.

[58] Whilst discussing a similar issue, Harry Haskell records a comment once made by a colleague of musicologist Hugo Leichtentritt, 'that in contrast to French musicians, who normally performed in modern concert attire, the Germans wore costumes that "took the taste of the eighteenth century into account, although the men could not bring themselves to sacrifice their moustaches in the cause of stylistic propriety" ' (*The Early Music Revival*, 55).

practices and concepts that existed only long ago. The ideal has too often functioned, in their view, with the naïve belief that a pure understanding of the past can be obtained untainted by an understanding formed by the present. As long as these beliefs are held, they say, the ideal of authenticity, employed to justify the performance of early music on original instruments, operates fallaciously.

By way of reaction, some defenders of the movement have shown willingness to dispense with the notion of authenticity, choosing, instead, to concentrate on the historically informed nature of their enterprise. One can, they say, perform in an historically informed manner, without that entailing that one has replicated perfectly the conditions of the past. By performing in an informed way, one just comes a little closer to what the past must have been like.

Others have proposed a notion of authenticity minus the historicist bent. One needs to be faithful more to the music itself than to all the conditions of the original historical context. One can distinguish being true to the music from being true to history. Though the idea of being true to the music might provide musicians with a novel way to decide what is and what is not essential to any given musical composition, this idea has come dangerously close to being anachronistically employed. For the ideal of being true to music has all too often been conflated with the ideal of being true to the work. What, after all, is the result of conceiving of early music in isolation or selective isolation from the original context of performance, if not a self-sufficent musical work? This conflation is not a necessary one, but it has come about through the influence of a pervasive aesthetic and range of concepts which allow one so very easily to think about early music in terms provided by the *Werktreue* ideal.[59]

Given the current terms of the debate, it remains easier, in fact, to distinguish between two rather than three ideals of authentic performance. Though neither mutually exclusive nor unproblematic in themselves, one can differentiate authentic performance guided by

---

[59] A similar argument was recently offered by Donald Henehan, who, convinced by an argument of Richard Taruskin, wrote: 'The fortepianist . . . is not reproducing what the 18th-century listener might recognize as Mozart's emotionally powerful music but remodeling Mozart in line with modern esthetics: concern for clarity, lightness, irony, . . . and, in Mr. Taruskin's words, "the ideology of our museum culture." All music composed or played in our century pays homage in greater or lesser degree to such ideals, which are so pervasive that we are barely, if at all, conscious of them. And so, it should not be surprising that performers honor their own time, consciously or not, whether the music under their hands is old or new.' ('Early Music is as Old as Today', *New York Times*, Arts Section (3 Sept. 1989), 17.)

Werktreue, from that guided by the ideal of being historically informed. Using the former ideal, one strives at least to comply perfectly with the public expression of the composer's intentions, perhaps the score, perhaps even a modernized score, if one believes that that more adequately than the original score represents the work. Using the latter ideal, one strives to reproduce as much of the original historical setting as possible, so as to be true, as far as one can be, to the original compositional and performance practices, so that one will come in turn a little closer to how the music once sounded. In both cases, one aims to reproduce in performance the 'true' meaning of the music.

Neither of these two ideals, however, tells us enough about the authenticity of early music performances. As long as it carries the baggage of a romantic aesthetic, the *Werktreue* ideal is at best being employed anachronistically. The ideal of historically informed fidelity, as defined above, is too general and applies equally to the performance of music of every sort. To continue to employ either of these ideals in the performance of early music still begs, then, for a satisfactory defence. What seems to be needed at this point is a clearer understanding of early music that renders an ideal of authenticity, perhaps more than one, appropriate to the conditions of early music production. Thus the return of a distinct, third ideal.

A third ideal is needed, however, only if players of early music wish to continue to speak of their performances as authentic or as being true to something. One could, perhaps, stop debating the merits of performances on the grounds of their being more or less authentic. One could cease debating their merits in these terms for the same sorts of reasons that one might decide to stop judging performances of concert music as being more or less true to the relevant works, as if the works were fixed once and for all in their meaning. This sort of strategy corresponds to what now might be called a 'post-modern' conception of performance and interpretation that invokes a denial of any commitment to fidelity, however fidelity is understood. Some theorists have already begun to move in this direction by speaking of a multiplicity of ways to be authentic, where each way depends on what one wants to be authentic to. But retaining the terminology of authenticity remains problematic while its use continues—as I believe it still does—to carry absolutist and unduly rigid, objectivist sentiments.

I am not yet convinced by the 'post-modern' move. I am not at all sure that we can shed ourselves completely of any commitment to 'authenticity', especially if we understand our 'being true to something' just to mean that we have chosen to be regulated by a given set of ideals. I am also not convinced that, as a post-modernist, one could justify rewriting the history of music produced under the work-concept, according to a non-work-based aesthetic, to make it look as if musicians had, despite their 'misleading' claims to the contrary, never 'really' composed, performed, and conducted works. This strategy might be thought to correspond precisely to one's rejecting certain traditional (romantic) ideals and aesthetic theories in the process of demystifying the past. But I do not see why this revisionist and imperialist strategy follows, or how this strategy could be made compatible with other post-modern inclinations that resist such strong revisionist tendencies. I am, however, much more certain of the advantages that result from our more modestly allowing different performance movements to develop for themselves their own regulative ideals. (Of course, that does not mean that all will or should succeed.)

Thus, my conclusion. More than any other movement currently existing within the European tradition of classical music, the early music movement is perfectly positioned to present itself not only as a 'different way of thinking about music', but also as an alternative to a performance practice governed by the work-concept. By positioning itself as a viable and dynamic alternative, even as a challenge to another practice, it is able to serve as a constant and living reminder to all musicians that the *Werktreue* ideal can be delimited in scope. Such delimitation, resulting from the presentation and acceptance of viable alternative practices, is a very good thing. It keeps our eyes open to the possibility of producing music in new ways under the regulation of new ideals. It keeps our eyes open to the inherently critical and revisable nature of our regulative concepts. Most importantly, it helps us overcome that deep-rooted desire to hold the most dangerous of beliefs, that we have at any time got our practices absolutely right.

X

It has not been the primary intention of this book either to justify or dismantle, in aesthetic, historical, or philosophical terms, the existence of the regulative concept of a musical work. I have sought, first and foremost, for an adequate way to speak philosophically about musical works, in the hope that I would highlight certain facts about a practice that have long been hidden from our view. Without doubt, the most important fact is that being true to music or a particular type of music does not necessarily mean being true to a work. This lesson by itself is of substantial philosophical and musical significance.

This lesson points, first, to certain facts about cultural practices that philosophers should not ignore. Not all music, for example, is packaged in the same way, not all music falls under a given range of concepts in the same way, and how we think about different sorts of music is deeply historicized. A philosophical method that gives priority to these sorts of facts will employ an ontological model which needs and demands no break from the empirical, historical condition of phenomena. This is not the case with all philosophical methods. The analytic method, for example, places its interests and priorities elsewhere. As long as it continues to do so, it will find itself unable to account satisfactorily for the phenomena. As I have argued, it is not that analytic theorists have made false claims. The problem has resided, rather, in the status and formulation of their theoretical claims which, being unduly stringent in complex and subtle ways, have left the connection between claim and phenomena—theory and practice—more disconnected than connected.

The significance to the musical world of the lesson is rather different. It turns on those questions with which this book began: what is involved in the composition, performance, and reception of classical music? What are we doing when we listen to this music seriously? Why, when playing a Beethoven Sonata, do performers begin with the first note indicated in the score? Why don't they feel free to improvise around the Sonata's central theme? Why, finally, does it go against tradition for an audience at a concert of classical music to tap its feet? Now that we know that it is our commitment to the *Werktreue* ideal which is responsible for these behavioural patterns, the next question immediately arises: to what degree and at what cost have musicians been constrained by this commitment, or has it worked to every musician's advantage? Perhaps listening to the music produced in the last two centuries will prompt the 'easy' answer that without the constraint of this ideal, composition of the great symphonies, concertos, and sonatas would have been impossible. By itself, however, this could only ever be part of an answer, whatever romantics say about music's ability to take us beyond worldly concerns. Music as an end could never, on aesthetic grounds alone, fully justify the social or political means involved in its composition, performance, and reception. The question, therefore, still asks for a more satisfactory answer, one that will force us to think about music, less as excused and separated, and more as inextricably connected to the ordinary and impure condition of our human affairs. The imaginary museum of musical works may well remain imaginary, as it continues to display the temporal art of music in the plastic terms of works of fine art, but it will never achieve complete transcendence and purity while it allows human beings to enter through its doors.

BIBLIOGRAPHY OF WORKS CITED

ABRAHAM, G., *The Concise Oxford History of Music* (Oxford, 1979).
—— (ed.), *The New Oxford History of Music* (10 vols.; Oxford, 1986).
ABRAMS, M. H., *The Mirror and the Lamp: Romantic Theory and the Critical Tradition* (Oxford, 1953).
ADDISON, J., *et al.*, *The Spectator*, ed. G. A. Aitken (6 vols.; London, n.d.).
ADORNO, T. W., *Prisms*, tr. S. Weber and S. Weber (Cambridge, Mass., 1967).
—— *Philosophy of Modern Music*, tr. A. G. Mitchell and W. V. Blomster (New York, 1973).
—— *Introduction to the Sociology of Music*, tr. E. B. Ashton (New York, 1976).
—— *Aesthetic Theory*, tr. C. Lenhardt, ed. G. Adorno and R. Teidemann (London, 1984).
ALISON, A., *Essays on the Nature and Principles of Taste* (2 vols.; Edinburgh, 1815).
ALLEN, W. D., *Philosophies of Music History: A Study of General Histories of Music, 1600–1960* (New York, 1962).
ALPERSON, P. (ed.), *What is Music? An Introduction to the Philosophy of Music* (New York, 1987).
ANDERBERG, T., NILSTUN, T., and PERSSON, I. (eds.), *Aesthetic Distinction: Essays Presented to Göran Hermerèn* (Lund, 1988).
ANDERSON, J. C., 'Musical Identity', *Journal of Aesthetics and Art Criticism*, 40 (1982), 285–91.
—— 'Musical Kinds', *British Journal of Aesthetics*, 25 (1985), 43–9.
ARISTOTLE, *The Works of Aristotle*, ed. W. D. Ross (12 vols.; Oxford, 1908–52).
—— *The Basic Works of Aristotle*, ed. R. McKeon (New York, 1941).
ATKINS, H., and NEWMAN, A. (eds.), *Beecham Stories: Anecdotes, Sayings and Impressions of Sir Thomas Beecham* (New York, 1978).
ATTALI, J., *Noise: The Political Economy of Music*, tr. B. Massumi (Minneapolis, 1985).
AUERBACH, E., *Mimesis: The Representation of Reality in Western Literature*, tr. W. R. Trask (Princeton, 1973).
AUGUSTINE, *Confessions*, tr. R. S. Pine-Coffin (New York, 1961).
BABBITT, M., 'Who Cares if you Listen?' *High Fidelity* (1958), rpt. as 'The Composer as Specialist', in Kostelanetz (ed.) *Esthetics Contemporary*, 280–7.
BACH, C. Ph. E., *Versuch über die wahre Art das Klavier zu spielen* (Berlin, 1762).
BALLANTINE, C., 'Towards an Aesthetics of Experimental Music', *The Musical Quarterly*, 63 (1977), 224–46.

BARKER, A. (ed.), *Greek Musical Writings, i. The Musician and his Art* (Cambridge, 1984).

BARTHES, R., *The Rustle of Language*, tr. R. Howard (Berkeley/Los Angeles, 1986).

BARTÓK, B., 'Race Purity in Music', *Modern Music: A Quarterly Review*, 19 (1942), 153–5.

BARZUN, J., *Berlioz and his Century: An Introduction to the Age of Romanticism* (Chicago, 1982).

BATTEUX, C., *Les Beaux Arts réduits à un même principe* (Paris, 1746).

BEARDSLEY, M. C., *Aesthetics from Classical Greece to the Present: A Short History* (Univ. of Alabama, 1966).

BEETHOVEN, L. VAN *Beethoven: Letters, Journals and Conversations*, ed. M. Hamburger (New York, 1951).

—— *Letters of Beethoven*, ed. E. Anderson (3 vols.; New York, 1961).

BENT, I. D., *Analysis*, New Grove Handbooks in Music (New York/London, 1987/8).

BERLIOZ, H., *A travers chant: Études musicales: Adorations, boutades et critiques*, 2nd edn. (Paris, 1872).

—— *Symphonie fantastique* (1830) (Leipzig, 1900), 1.

BERNSTEIN, L., *The Joy of Music* (London, 1959).

BLUME, F., *Classic and Romantic Music: A Comprehensive Survey*, tr. M. D. Herter Norton (London, 1979).

BOTTOMORE, T., *The Frankfurt School* (London/New York, 1984).

BOURDIEU, P., *Distinction: A Social Critique of the Judgement of Taste*, tr. R. Nice (Cambridge, Mass., 1984).

BRADBURY, M., and MCFARLANE, J. (eds.), *Modernism, 1890–1930* (Harmondsworth, Middx., 1976).

BRENDEL, A., *Musical Thoughts and Afterthoughts* (Princeton, 1976).

BRODY, B. A., *Identity and Essence* (Princeton, 1980).

BROSSARD, S. DE, *Dictionaire de musique* (Paris, 1703).

BUJIĆ, B., (ed.), *Music in European Thought, 1851–1912* (Cambridge, 1988).

BUNGAY, S., *Beauty and Truth: A Study of Hegel's Aesthetics* (Oxford, 1984).

BURBIDGE, P., and SUTTON, R. (eds.), *The Wagner Companion* (London, 1979).

BÜRGER, P., *Theory of the Avant-garde*, tr. M. Shaw (Minneapolis, 1984).

BURKHARDT, H., and SMITH, B. (eds.), *Handbook of Metaphysics and Ontology* (2 vols.; Munich, 1991).

BURKHOLDER, J. P., 'Museum Pieces: The Historicist Mainstream in Music of the Last Hundred Years', *Journal of Musicology*, 2 (1983), 115–34.

BURNEY, C., *An Eighteenth-Century Musical Tour in France and Italy* (1773), ed. P. A. Scholes (Westport, Conn., 1979).

BUSCHEL, B., 'Angry Young Man with a Horn', *Gentleman's Quarterly*, 57 (1987), 192–5, 227–34.

CAGE, J., *Silence: Lectures and Writings* (Middletown, Conn., 1961).

CANETTI, E., *Crowds and Power*, tr. C. Stewart (New York, 1962).

CARPENTER, P., 'The Musical Object' ed. C. Seltzer, *Current Musicology*, Special Project, 5 (1967), 56–87.

CARRIER, D., 'Interpreting Musical Performances', *Monist*, 66 (1983), 202–12.

CARSE, A., *The Orchestra in the Eighteenth Century* (Cambridge, 1940).

—— *The History of Orchestration* (New York, 1964).

CHARLTON, W., *Aesthetics* (London, 1970).

CHAVEZ, C., *Toward a New Music: Music and Electricity*, tr. H. Weinstock (New York, 1975).

COHEN, A., *Music in the French Royal Academy of Sciences: A Study in the Evolution of Musical Thought* (Princeton, 1981).

COHEN, T., 'The Possibility of Art: Remarks on a Proposal by Dickie', *Philosophical Review*, 82 (1973), 69–82.

COLLES, H. C., 'Extemporization or Improvisation', in Grove, *Dictionary*, 5th edn., ii. 991–3

COLLINGWOOD, R. G., *The Principles of Art* (Oxford, 1938).

COMFORT, A., 'Art and Social Responsibility: The Ideology of Romanticism', in Gleckner and Enscoe, (eds.) *Romanticism*, 168–81.

CONE, E. T., 'One Hundred Metronomes', *American Scholar*, 40 (1977), 443–59.

CONSTANT, B. DE REBECQUE, *Journaux intimes* (n.p., 1952).

'Copyright', *Encyclopedia Britannica*, 11th. edn. (Cambridge, 1910), vii. 118–29.

COX, R., 'Are Musical Works Discovered?', *Journal of Aesthetics and Art Criticism*, 43 (1985), 367–74.

CROCE, B., *Aesthetic as Science of Expression and General Linguistic*, tr. D. Ainslie (New York, 1922).

CROFTON, I., and FRASER, D. (eds.), *A Dictionary of Musical Quotations* (New York, 1985).

DAHLHAUS, C., *Esthetics of Music*, tr. W. Austin (Cambridge, 1982).

—— *Foundations of Music History*, tr. J. B. Robinson (Cambridge, 1983).

—— *Schoenberg and the New Music*, tr. D. Puffett and A. Clayton (Cambridge, 1987).

—— *Nineteenth-Century Music*, tr. J. Bradford Robinson (Berkeley/Los Angeles, 1989).

—— *The Idea of Absolute Music*, tr. R. Lustig (Chicago, 1989).

DANCY, J., MORAVSCIK, J. M. E., and TAYLOR, C. C. W. (eds.), *Human Agency: Language, Duty, and Value. Philosophical Essays in Honour of J. O. Urmson* (Stanford, Calif., 1988).

DANIÉLOU, A., and BRUNET, J., *The Situation of Music and Musicians in Countries of the Orient*, tr. J. Evarts (Florence, 1971).

DANTE, *De vulgari eloquentia*, tr. A. G. Ferrers Howell (London, 1890).

DANTO, A., *The Transfiguration of the Commonplace: A Philosophy of Art* (Cambridge, Mass., 1981).

DAVID, H. T., and MENDEL, A. (eds.), *The Bach Reader: A Life of Johann Sebastian Bach in Letters and Documents* (New York, 1966).

DAVIES, S., 'Transcription, Authenticity and Performance', *British Journal of Aesthetics*, 28 (1988), 216–27.

DEBUSSY, C., *Debussy on Music*, ed. R. Langham Smith (Ithaca, 1977).

DEWEY, J., *The Quest for Certainty: A Study of the Relation of Knowledge and Action* (New York, 1929).

DICKIE, G., *The Art Circle: A Theory of Art* (New York, 1984).

DIFFEY, T. J., 'Essentialism and the Definition of "Art" ', *British Journal of Aesthetics*, 13 (1973), 103–20.

DIPERT, R. R., 'Types and Tokens: A Reply to Sharpe', *Mind*, 89 (1980), 587–8.

DONNINGTON, R., *Baroque Music: Style and Performance: A Handbook* (New York, 1982).

—— 'Ornamentation', in Grove, *Dictionary*, 5th edn., vi. 365–478.

DREYFUS, L., 'Early Music Defended against its Devotees: A Theory of Historical Performance in the Twentieth Century', *Musical Quarterly*, 69 (1983), 297–322.

DUCHAMP, M., *The Writings of Marcel Duchamp*, ed. M. Sanouillet and E. Peterson (Oxford, 1973).

DUFRENNE, M., *The Phenomenology of Aesthetic Experience*, tr. E. S. Casey *et al.* (Evanston, Ill., 1973).

DWORKIN, R. M. (ed.), *The Philosophy of Law* (Oxford, 1987).

EAGLETON, T., *Literary Theory: An Introduction* (Minneapolis, 1983).

—— *The Ideology of the Aesthetic* (Oxford, 1990).

FORKEL, J. N., *Johann Sebastian Bach: His Life, Art and Work* (1802), tr. C. S. Terry (London, 1920).

FOUCAULT, M., *The Archaeology of Knowledge and the Discourse on Language*, tr. A. M. Sheridan Smith (New York, 1972).

FRITH, S., 'Towards an Aesthetic of Popular Reception', in Leppert and McClary (eds.) *Music and Society*, 133–49.

GADAMER, H. G., *Truth and Method* (London, 1979).

GALKIN, E. W., *A History of Orchestral Conducting: In Theory and Practice* (New York, 1988).

GALLIE, W. B., *Philosophy and the Historical Understanding* (London, 1964).

GAY, P., *The Enlightenment: An Interpretation, The Science of Freedom* (New York, 1969).

GEIRINGER, K., *Haydn: A Creative Life in Music* (New York, 1946).

GEUSS, R., *The Idea of a Critical Theory: Habermas and the Frankfurt School* (Cambridge, 1981).

GLECKNER, R. F., and ENSCOE, G. E. (eds.), *Romanticism: Points of View* (Englewood Cliffs, NJ, 1962).

GOEHR, L., 'Being True to the Work', *Journal of Aesthetics and Art Criticism*, 47 (1989), 56–67.

—— Review of J. J. Winkelmann's *Reflections on the Imitation of Greek Works in Painting and Sculpture*, tr. E. Heyer and R. C. Norton, in *Teaching Philosophy*, 12: 3 (1989), 329–32.

—— 'The Power of the Podium', *Yale Review*, 79: 3 (1990), 365–81.

—— 'Concepts, open', in Burkhardt and Smith (eds.), *Handbook of Metaphysics and Ontology*, i. 166–7.

—— 'Music Has no Meaning to Speak of: On the Politics of Musical Interpretation', in M. Krausz (ed.), *The Interpretation of Music* (Oxford, forthcoming).

GOETHE, J. W. VON, *Elective Affinities*, tr. R. J. Hollingdale (Harmondsworth, Middx., 1971).

GOMBRICH, E. H., *In Search of Cultural History* (Oxford, 1969).

GOODMAN, N., *The Structure of Appearance* (Cambridge, Mass., 1951).

—— *Languages of Art: An Approach to a Theory of Symbols* (Oxford, 1969).

—— *Ways of Worldmaking* (Hassocks, Sussex, 1978).

GRAF, P., and KRZEMIEN-OJAK, S. (eds. and trs.), *Roman Ingarden and Contemporary Polish Aesthetics* (Warsaw, 1975).

GRANGE, H.-L. DE LA, *Mahler*, i (New York, 1973).

GRASSINEAU, J., *A Musical Dictionary: A Facsimile of the 1740 London Edition* (New York, 1966).

GREEN, L., *Music on Deaf Ears: Musical Meaning, Ideology and Education* (Manchester, 1988).

GRIFFITHS, P., *A Concise History of Avant-garde Music: From Debussy to Boulez* (New York, 1978).

GRILLPARZER, F., 'In Moscheles Stammbuch' (1826), *Sämtliche Werke: Gedichte*, iii. 1 (Vienna, 1937).

GROVE, G., *A Dictionary of Music and Musicians (A.D. 1450–1889) by Eminent Writers, English and Foreign* (London, 1879–89); also 5th edn., ed. E. Blom (9 vols.; New York, 1954).

HANDEL, G. F., *The Letters and Writings of Georg Frideric Handel*, ed. E. H. Müller (New York, 1935).

HANSLICK, E., *On the Musically Beautiful*, tr. and ed. G. Payzant (Indianapolis, 1986).

HARDING, J., *Gounod* (New York, 1973).

HARRIS, J., *Three Treatises: The First concerning Art, the Second concerning Music, Painting, and Poetry, the Third concerning Happiness* (London, 1744).

HARRISON, N., 'Types, Tokens and the Identity of the Musical Work', *British Journal of Aesthetics*, 15 (1975), 336–46.

HART, H. L. A., 'The Ascription of Responsibility and Rights', *Proceedings of the Aristotelian Society*, 49 (1948–9), 179–94.

HASKELL, H., *The Early Music Revival: A History* (London, 1988).

HAYDN, F. J., *The Collected Correspondence and London Notebooks of Joseph Haydn*, ed. H. C. Robbins Landon (London, 1959).

HEGEL G. W. F., *Philosophy of Right*, tr. T. M. Knox (Oxford, 1967).

—— *Aesthetics: Lectures on Fine Art*, tr. T. M. Knox (2 vols.; Oxford, 1975).

—— *Phenomenology of Spirit*, tr. A. Miller and J. N. Findlay (Oxford, 1977).

HEIDEGGER, M., *Poetry, Language, Thought*, tr. A. Hofstadter (New York, 1971).

HENEHAN, D., 'Early Music is as Old as Today', *New York Times*, Arts Section (3 Sept. 1989), 17.

HERON-ALLEN, E., 'Tartini', in Grove *Dictionary*, 5th edn., viii. 312–15.

HIGGINS, D., 'Intermedia', in Kostelanetz (ed.), *Esthetics Contemporary*, 186–90.

HOBSBAWM, E. J., Review of *Duke Ellington* by J. L. Collier, *New York Review of Books* (19 Nov. 1987), 3–7.

HOFFMAN, R., 'Conjectures and Refutations on the Ontological Status of the Work of Art', *Mind*, 71 (1962), 512–20.

HOFFMANN, E. T. A., *Musikalische Novellen und Aufsätze*, i, ed. E. Istel (Regensburg, 1919).

HOGARTH, G., *Music History, Biography and Criticism* (New York, 1845).

HOGWOOD, C., *Handel* (New York, 1984).

HOROWITZ, J., *Understanding Toscanini: How he Became an American Culture-God and Helped Create a New Audience for Old Music* (Minneapolis, 1987).

HOSLER, B., *Changing Aesthetic Views of Instrumental Music in Eighteenth Century Germany* (Ann Arbor, Mich., 1981).

HOUGH, G., 'The Modernist Lyric', in Bradbury and McFarlane (eds.), *Modernism*, 312–22.

HOY, D. C., *The Critical Circle: Literature, History, and Philosophical Hermeneutics* (Berkeley/Los Angeles, 1982).

HUSSERL, E., *Experience and Judgment: Investigations in a Genealogy of Logic*, ed. L. Landgrebe (London, 1973).

HYATT, A., 'General Musical Conditions', in Abraham (ed.), *The New Oxford History of Music*, viii, 1–25.

INGARDEN, R., *Time and Modes of Being*, tr. H. R. Michejda (Springfield, Ill., 1964).

—— 'Bemerkungen zu den Bemerkungen von Professor Zofia Lissa', *Studia Filozoficzne*, 4 (1970), 351–63.

—— *The Work of Music and the Problem of its Identity* (1928), tr. A. Czerniawski, ed. J. G. Harrell (Berkeley/Los Angeles, 1986).

IVES, C., *Essays before a Sonata and Other Writings* (New York, 1961).

JACOBSON, R., *Reverberations: Interviews with the World's Leading Musicians* (New York, 1974).

JAMES, F. E. SKONE, 'Copyright', in Grove *Dictionary*, 5th edn., ii. 430–4.

JANDER, O., 'Putting the Program back in Program Music', *New York Times*, Arts Section (3 Sept. 1989), 17.

JAUSS, H. R., *Toward an Aesthetic of Reception*, tr. T. Bahti (Minneapolis, 1982).

KANT, I., *Critique of Pure Reason*, tr. N. Kemp Smith (New York, 1929).
—— *Critique of Judgement*, tr. J. C. Meredith (Oxford, 1952).

KEATES, J., Review of Reinhard Strohm's *Essays on Handel and Italian Opera*, *Times Literary Supplement* (31 Jan. 1986), 118.

KEATS, J., *Letters of John Keats*, ed. M. B. Forman (Oxford, 1952).

KELLER, H., *Criticism*, ed. J. Hogg (London, 1987).

KENDALL, A., *The Tender Tyrant: Nadia Boulanger* (Wilton, Conn., 1976).

KENNICK, W. E., 'Does Traditional Aesthetics Rest on a Mistake?', *Mind*, 67 (1958), 317–34.

KENYON, N. (ed.), *Authenticity and Early Music: A Symposium* (Oxford, 1988).

KERMAN, J., *Contemplating Music: Challenges to Musicology* (Cambridge, Mass., 1985).

KIERKEGAARD, S., *Either/Or*, tr. D. F. Swenson and L. M. Swenson (Princeton, 1944).

KIVY, P., 'Mainwaring's Handel: Its Relation to English Aesthetics', *Journal of the American Musicological Society*, 17 (1964), 170–8.
—— *The Corded Shell: Reflections on Musical Expression* (Princeton, 1980).
—— 'Platonism in Music: A Kind of Defense', *Grazer Philosophische Studien*, 19 (1983), 109–29.
—— *Osmin's Rage: Philosophical Reflections on Opera, Drama, and Text* (Princeton, 1988).

KOSTELANETZ, R., *On Innovative Musicians* (New York, 1989).
—— (ed.), *Esthetics Contemporary* (Buffalo, NY., 1978).

KRAUS, R. C., *Pianos and Politics in China: Middle-Class Ambitions and the Struggle over Western Culture* (New York, 1989).

KRAUSZ, M. (ed.), *The Interpretation of Music* (Oxford, forthcoming).

KRIPKE, S. A., *Naming and Necessity* (Oxford, 1980).

KRISTELLER, P. O., 'The Modern System of the Arts', rpt. in Weitz (ed.), *Problems in Aesthetics*, 108–64.

LAKATOS, I., *Mathematics, Science, and Epistemology*, ed. J. Worrall and G. Currie (London/New York, 1978).

LANG, B. (ed.), *The Concept of Style* (Philadelphia, 1979).

LE HURAY, P., and DAY, J. (eds.), *Music and Aesthetics in the Eighteenth and Early-Nineteenth Centuries* (Cambridge, 1981).

LEPPERT, R., and McCLARY, S. (eds.), *Music and Society: The Politics of Composition, Performance and Reception* (Cambridge, 1987).

LEPPMANN, W., *Winckelmann* (New York, 1970).

LESSING, G.E., 'Laöcoon or on the Limits of Painting and Poetry', in Nisbet (ed.), *German Aesthetic and Literary Criticism*, 55–134.

LEVINE, L. W., *Highbrow/Lowbrow: The Emergence of Cultural Hierarchy in America* (Cambridge, Mass., 1988).

LEVINSON, J., 'Autographic and Allographic Art Revisited', *Philosophical Studies*, 38 (1980), 367–83.

—— 'What a Musical Work is', *Journal of Philosophy*, 77 (1980), 5–28.

—— 'Artworks and their Futures', in Anderberg *et al.* (eds.), *Aesthetic Distinction*, 56–84.

—— *Music, Art and Metaphysics* (Ithaca, 1990).

LEWIS, C. I., *An Analysis of Knowledge and Valuation* (La Salle, Ill., 1946).

LISSA, Z., 'Some Remarks on the Ingardenian Theory of the Musical Work', in Graf and Krzemien-Ojak, (eds.), *Roman Ingarden*, 129–44.

LISTENIUS, N., *Musica: Ab authore denuo recognita multisque novis regulis et exemplis adaucta* (1549); facsimile, ed. G. Schünemann (Berlin, 1927).

LISZT, F., 'A Letter on Conducting' (1853), *Gesammelte Schriften*, ed. L. Ramann, v (Leipzig, 1880–3), 229–32.

LORD, C., 'A Kripkean Approach to the Identity of a Work of Art', *Journal of Aesthetics and Art Criticism*, 35 (1977), 147–55.

MACHLIS, J., *Introduction to Contemporary Music* (New York, 1961).

MALRAUX, A., *The Voices of Silence*, tr. S. Gilbert (Princeton, 1978).

MANDELBAUM, M., 'Family Resemblances and Generalizations concerning the Arts', *American Philosophical Quarterly*, 2 (1965), 219–28.

MARGOLIS, J., 'Works of Art as Physically Embodied and Culturally Emergent Entities', *British Journal of Aesthetics*, 14 (1974), 187–96.

—— (ed.), *Philosophy Looks at the Arts: Contemporary Readings in Aesthetics*, 3rd edn. (Philadelphia, 1987).

MELLORS, W., *François Couperin and the French Classical Tradition* (New York, 1968).

MERRICK, P., *Revolution and Religion in the Music of Liszt* (Cambridge, 1987).

MERTENS, W., *American Minimal Music* (London/White Plains, NY, 1983).

MEYER, L. B., *Style and Music: Theory, History, and Ideology* (Philadelphia, 1989).

MILL, J. S., *Utilitarianism, On Liberty and Considerations of Representative Government*, ed. H. B. Acton (London, 1972).

MILLER, J., 'Bird Lives in his Incandescent Art', *Newsweek* (31 Oct. 1988), 71.

MONTEVERDI, C., *The Letters of Claudio Monteverdi*, tr. D. Stevens (Cambridge, 1980).

MORROW, M., 'Musical Performance', *Early Music*, 6 (1978), 233–46.

MOZART, L., *Versuch einer gründlichen Violinschule* (Augsburg/Vienna, 1756).

MOZART, W. A., *The Letters of Mozart and his Family*, ed. E. Anderson (3 vols.; London, 1938).

—— *Mozart's Letters*, ed. E. Blom (Harmondsworth, Middx., 1956).

NETTL, B., *The Study of Ethnomusicology: Twenty-nine Issues and Concepts* (Urbana/Chicago, 1983).

NEUBAUER, J., *The Emancipation of Music from Language: Departure from Mimesis in Eighteenth-Century Aesthetics* (New Haven, Conn., 1986).

NEWMAN, W. S., 'Emanuel Bach's Autobiography', *Musical Quarterly*, 51 (1965), 363–86.

NIETZSCHE, F. W., *Basic Writings of Nietzsche*, tr. and ed. W. Kaufmann (New York, 1968).

—— *Twilight of the Idols*, tr. R. J. Hollingdale (Harmondsworth, Middx., 1968).

NISBET, H. B. (ed.), *German Aesthetic and Literary Criticism: Winckelmann, Lessing, Hamann, Herder, Schiller and Goethe* (Cambridge, 1985).

NORRIS, C. (ed.), *Music and the Politics of Culture* (London, 1989).

NORTH, R., *Roger North on Music: Being a Selection from his Essays Written during the Years* c.1695–1728, ed. J. Wilson (London, 1959).

OGDEN, C. K. (ed.), *Bentham's Theory of Fictions* (New York, 1932).

ONG, W. J., *The Presence of the Word: Some Prolegomena for Cultural and Religious History* (Minneapolis, 1967).

PALMER, R. E., *Hermeneutics* (Evanston, Ill., 1969).

PANOWSKY, E., 'The Ideological Antecedents of the Rolls Royce Radiator', *Proceedings of the American Philosophical Society*, 107 (1963), 273–88.

PARFIT, D., *Reasons and Persons* (Oxford, 1984).

PATER, W., *The Renaissance: Studies in Art and Poetry*, ed. A. Phillips (Oxford, 1986).

PEACHAM, H., *The Complete Gentleman: The Truth of our Times and the Art of Living in London*, ed. V. B. Heltzel (Ithaca, 1962).

PLATO, *The Laws*, tr. T. J. Saunders (Harmondsworth, Middx., 1970).

—— *Plato's Republic*, tr. G. M. A. Grube (Indianapolis, 1974).

POGGIOLI, R., *The Theory of the Avant-garde*, tr. G. Fitzgerald (Cambridge, Mass., 1968).

POLANYI, M., *Personal Knowledge* (London, 1962).

PRENDERGAST, R. M., *Film Music: A Neglected Art* (New York, 1977).

PRICE, K., 'What is a Piece of Music?', *British Journal of Aesthetics*, 22 (1982), 322–36.

PUTNAM, H., *Collected Papers*, ii (Cambridge, 1975).

QUANTZ, J. J., *On Playing the Flute* (1752), tr. E. R. Reilly (New York, 1985).

RAMEAU, J. P., *Treatise on Harmony* (1722), tr. P. Gossett (New York, 1971).

RAWLS, J., 'Two Concepts of Rules', *Philosophical Review*, 64 (1955), 3–32.

RAYNOR, H., *Music and Society since 1815* (London, 1976).

REYNOLDS, J., *Discourses on Art* (New York, 1961).

RICHARDSON, J., *A Discourse on the Dignity, Certainty, Pleasure and Advantage of the Science of the Connoisseur: Two Discourses* (London, 1719).

ROCKWELL, J., *All American Music: Composition in the Late Twentieth Century* (New York, 1983).

—— Review, *New York Times* (25 June 1989), § 2, 27, and 42.

ROSEN, C., 'Should Music be Played "Wrong"?', *High Fidelity*, 21 (1971), 54–8.

—— *The Classical Style: Haydn, Mozart and Beethoven* (New York/London, 1972).

ROUSSEAU, J.-J., *Œuvres complètes de J. J. Rousseau*, 'Beaux-Arts: Dictionnaire de Musique', tr. and ed. V. D. Musset-Pathay (2 vols.; Paris, 1824).

—— *Politics and the Arts*, 'Letter to d'Alembert on the Theatre', tr. A. Bloom (Ithaca, 1960).

ROWELL, L., *Thinking about Music: An Introduction to the Philosophy of Music* (Amherst, Mass., 1983).

SALMEN, W. (ed.), *The Social Status of the Professional Musician from the Middle Ages to the Nineteenth Century*, tr. H. Kaufman and B. Reisner (New York, 1983).

SALMON, N., *Reference and Essence* (Oxford, 1982).

SARTRE, J.-P., *The Psychology of Imagination*, tr. anon. (New York, 1965).

SAVILE, A., *The Test of Time: An Essay in Philosophical Aesthetics* (Oxford, 1982).

SCHAFER, R. MURRAY, *E. T. A. Hoffmann and Music* (Toronto, 1975).

SCHILLER, F., 'On Naive and Sentimental Poetry', in Nisbet (ed.), *German Aesthetic and Literary Criticism*, 177–232.

SCHLEGEL, F. VON, 'Athenäums-Fragmente', *Kritische und theoretische Schriften* (Stuttgart, 1978).

SCHOENBERG, A., *Letters*, ed. E. Stein (London, 1964).

—— *Style and Idea: Selected Writings of Arnold Schoenberg*, ed. L. Stein, tr. L. Black (Berkeley/Los Angeles, 1975).

SCHOPENHAUER, A., *The World as Will and Representation*, tr. E. F. J. Payne (2 vols.; New York, 1969).

SCHUELLER, H. M., *The Idea of Music: An Introduction to Musical Aesthetics in Antiquity and the Middle Ages* (Kalamazoo, Mich., 1988).

SCHUMANN, R., *Schumann on Music: A Selection from the Writings*, ed. H. Pleasants (New York, 1965).

SCHWARTZ, S. P. (ed.), *Naming, Necessity, and Natural Kinds* (Ithaca, 1977).

SCRUTON, R., 'Programme Music', *The New Grove Dictionary*, xv (London, 1981), 283–7.

—— *The Aesthetic Understanding: Essays in the Philosophy of Art and Culture* (London, 1983).

SEIDEL, W., *Werk und Werkbegriff in der Musikgeschichte* (Darmstadt, 1987).

SENECA, *La lettera 65 di Seneca*, ed. G. Scarpat (Brescia, 1970).

SENNETT, R., *The Fall of Public Man* (New York, 1977).

SHARPE, R. A., 'Type, Token, Interpretation and Performance', *Mind*, 88 (1979), 437–40.

SHATTUCK, R., 'Most of Modern Music', *New York Review of Books* (15 Mar. 1990), 32–3.

SHAW, G. BERNARD, *London Music in 1888–9 as heard by Corno di Bassetto (Later Known as Bernard Shaw) with Some Further Autobiographical Particulars* (London, 1937).

SHUSTERMAN, R. (ed.), 'Analytic Aesthetics', *Journal of Aesthetics and Art Criticism*, 46, Special Issue (1987).

SLOMINSKY, N., *Lexicon of Musical Invective: Critical Assaults on Composers since Beethoven's Time* (Seattle, 1965).

SMITH, A., *The Works of Adam Smith* (5 vols.; London, 1811).

SONNECK, O. G. (ed.), *Beethoven: Impressions by his Contemporaries* (New York, 1954).

SPARSHOTT, F. E., 'Aesthetics of Music', *The New Grove Dictionary*, i (London, 1981), 120–34.

—— 'Aesthetics of Music: Limits and Grounds', in Alperson (ed.) *What is Music?*, 33–98.

SPOHR, L., *Autobiography*, tr. anon. (London, 1865).

STENDHAL, *The Life of Rossini*, tr. R. N. Coe (London/New York, 1985).

STRAVINSKY, I., and CRAFT, R., *Conversations with Igor Stravinsky* (Garden City, NY, 1959).

STRAWSON, P. F., 'Aesthetic Appraisal and Works of Art', *Freedom and Resentment and Other Essays* (London, 1974), 178–88.

STRUNK, O., *Source Readings in Music History: From Classical Antiquity through the Romantic Era* (New York, 1950).

TANNER, M., 'The Total Work of Art' in Burbidge and Sutton (eds.), *The Wagner Companion*, 140–224.

TARUSKIN, R., *et al.*, 'The Limits of Authenticity: A Discussion', *Early Music*, 12 (1984), 3–25, 523–5.

TATARKIEWICZ, W., *History of Aesthetics*, ed. J. Harrell (3 vols.; Warsaw, 1970).

TEMPLIER, P. D., *Erik Satie*, tr. E. L. French and D. S. French (Cambridge, Mass., 1969).

TOMKINS, C., 'Onward and Upward with the Arts', *The New Yorker* (24 Oct. 1988), 110–20.

TORMEY, A., 'Indeterminacy and Identity in Art', *Monist*, 58 (1974), 203–15.

TREITLER, L., *Music and the Historical Imagination* (Cambridge, Mass., 1989).

VOLTAIRE, *Candide*, tr. J. Butt (Harmondsworth, Middx., 1947).

WAISMANN, F., 'Verifiability', *Aristotelian Society Proceedings*, supp. vol. 19 (1945), 119–50.

WALKER, A., *Franz Liszt: The Virtuoso Years, 1811–1847* (Ithaca, 1987).

WALTER, B., *Gustav Mahler*, tr. J. Galton (New York, 1970).

WALTHER, J. G., *Musicalisches Lexicon* (Leipzig, 1732).

WALTON, K. L., 'Categories of Art', *Philosophical Review*, 66 (1970), 334–67.

—— 'Style and the Products and Processes of Art', in Lang (ed.), *The Concept of Style*, 45–66.

—— 'The Presentation and Portrayal of Sound Patterns', in Dancy *et al.* (eds.), *Human Agency*, 237–57, 301–2.

WATROUS, P., Article on Jazz, *New York Times* (10 Mar. 1990), Arts Section, 14.

WEBER C. M. VON, *Writings on Music*, tr. M. Cooper, ed. J. Warrack (Cambridge, 1981).

WEBER, W., 'Mass Culture and the Reshaping of European Musical Taste, 1770–1870', *International Review of the Aesthetics and Sociology of Music*, 8 (1977), 5–22.

WEBSTER, W. E., 'Music is not a Notational System', *Journal of Aesthetics and Art Criticism*, 29 (1971), 489–97.

—— 'A Theory of the Compositional Work of Music', *Journal of Aesthetics and Art Criticism*, 33 (1974), 59–66.

WEISS, P., and TARUSKIN, R., *Music in the Western World: A History of Documents* (New York, 1984).

WEITZ, M., 'The Role of Theory in Aesthetics', *Journal of Aesthetics and Art Criticism*, 15 (1956), 27–35.

—— (ed.), *Problems in Aesthetics: An Introductory Book of Readings* (New York/London, 1959).

WELLEK, R., and WARREN, A., *Theory of Literature* (New York/London, 1977).

WESLEY, S., *The Wesley Bach Letters: A facsimile Reprint of the First Printed Edition*, ed. P. Williams (London, 1988).

WICKE, P., *Rock Music: Culture, Aesthetics and Sociology* (Cambridge, 1990).

WILLHEIM, I., 'Johann Adolf Scheibe: German Musical Thought in Transition', diss. (Univ. of Illinois, 1963).

WILLIAMS, R., *Culture and Society 1780–1950* (Harmondsworth, Middx., 1963).

—— *The Politics of Modernism: Against the New Conformists*, ed. T. Pinkney (London/New York, 1989).

WINCKELMANN, J. J., *Reflections on the Imitation of Greek Works in Painting and Sculpture* (1755), tr. E. Heyer and R. C. Norton (La Salle, Ill. 1987).

WINN, J. A., *Unsuspected Eloquence: A History of the Relations between Poetry and Music* (New Haven, Conn., 1981).

WIORA, W., *Das musikalische Kunstwerk* (Tutzing, 1983).

WITTGENSTEIN, L., *Philosophical Investigations*, tr. G. E. M. Anscombe (Oxford, 1958).

WOLLHEIM, R., *Art and its Objects* (Harmondsworth, Middx., 1968).

——, WIGGINS, D., and GOODMAN, N., 'Are the Criteria of Identity that Hold for a Work of Art in the Different Arts Aesthetically Relevant?', *Ratio*, 19–20 (1978), 29–68.

WOLTERSTORFF, N., 'Towards an Ontology of Artworks', *Noûs*, 9 (1975), 115–42.

—— *Works and Worlds of Art* (Oxford, 1980).

—— 'The Work of Making a Work of Music', in Alperson (ed.) *What is Music?*, 102–29.

WÖRNER, K. H., *Stockhausen: Life and Work*, tr. B. Hopkins (Berkeley, Calif., 1973).

WUTHNOW, R., HUNTER, J. D., BERGESEN, A., and KURZWEIL, E., *Cultural Analysis: The Work of Peter L. Berger, Mary Douglas, Michel Foucault, and Jürgen Habermas* (London/Boston, 1984).

YOUNG-BRUEHL, E., *Hannah Arendt: For the Love of the World* (New Haven, Conn., 1982).

ZIFF, P., 'The Task of Defining a Work of Art', *Philosophical Review*, 63 (1953), 68–78.

—— 'Goodman's *Languages of Art*', *Philosophical Review*, 80 (1971), 509–15.

——, SIRCELLO, G., and WALTON, K. L., 'The Cow on the Roof', and replies in 'The Aesthetics of Music', *Journal of Philosophy*, 70 (1973), 713–26.

INDEX